Denise Maria Camargo Machado
José Leomar Fernandes Jr.

Standards for identifying defects in flexible pavements

Denise Maria Camargo Machado
José Leomar Fernandes Jr.

Standards for identifying defects in flexible pavements

Comparing the United States, France and Brazil

ScienciaScripts

Cover image: www.ingimage.com

This book is a translation from the original published under ISBN 978-3-330-76437-8.

Publisher:
Sciencia Scripts
is a trademark of
Dodo Books Indian Ocean Ltd. and OmniScriptum S.R.L publishing group

120 High Road, East Finchley, London, N2 9ED, United Kingdom
Str. Armeneasca 28/1, office 1, Chisinau MD-2012, Republic of Moldova, Europe
Managing Directors: Ieva Konstantinova, Victoria Ursu
info@omniscriptum.com

Printed at: see last page
ISBN: 978-620-8-50632-2

I dedicate this work to my parents in recognition of their support at every moment of my life, to my children as an incentive for their future and to my partner for his constant affection.

ACKNOWLEDGEMENTS

I thank God for the gifts he has given me, especially for the ability to never let myself down and always rebuild myself.

I would like to say a huge thank you to my supervisor, José Leomar, who always took an optimistic view of the difficult situations that arose over the course of my Master's degree, and knew how to encourage me to overcome them. On several occasions, I would arrive at our work meetings a little confused and full of doubts, but gradually everything became clear and I saw my strength and energy renewed to keep going.

I would like to thank my parents, Neusa and Orlando, for their love and unconditional dedication. Especially to my mother, who helped me finalise this text with her knowledge as a teacher. Thank you to my children, Tiago and Yara, a source of inspiration so that I don't give up on adding something to a world in need of good initiatives. Thank you to my partner, Paulo, who enlightened me with his optimism at times when discouragement came my way, and who also made several suggestions to contribute to my work.

I would like to thank my brothers, Antonio Carlos, Rogério and Ricardo, my sisters-in-law, Valéria and Franceli, as well as my nephews, Mariana, Rodrigo (my particular counsellor), Fernando and Lorena, who were all so present and intense in supporting and cheering me up, even during the time I was working on this dissertation. I would also like to thank my cousin Valéria and my friend Cláudia, who have been great supporters of my Master's from the very beginning.

I would like to thank my professional colleagues at the Department of Highways, who helped me throughout the period when I was involved in the research for my dissertation. In particular: Heloísa, with whom I often exchanged ideas about my work; Alfredo and Gilberto, who provided me with a lot of material from their own bibliographic collections; Toyama, who provided me with information for the Case Study; and my colleagues and bosses, Armando, Nakage and Domingos, who gave me support and advice whenever I needed it.

I would like to thank all the professors at USP, São Carlos Campus, for the lessons they taught me during my time at this institution. I often felt sincere understanding from the teachers and staff regarding my difficulties in doing the work in the same way as most of the students, due to the fact that I worked eight hours a day and had more limited time for the Master's course.

I would like to thank all the staff. In particular, Alexandre and Magali, who were always very attentive, and Beth and Heloísa, who helped me a lot throughout my Master's degree, with regularising documents, renewing enrolments, contacting professors and providing materials. Without this help, everything would have been much more difficult for me.

I would like to thank Mada, Diego and Davi for lending me a corner of their house to stay in once a week for almost a year. I would like to express my sincere thanks to all the dear colleagues who lived with me during my hectic Master's years, who became friends and companions for a period of time that will be unforgettable for me. Those were some very rich, happy and cosy days together. I hope I've added at least a little to each one's life, as much as each one has added to mine.

THANK YOU SO MUCH!

Giving up... I've thought about it seriously, but I've never really taken myself seriously; it's just that there's more ground in my eyes than tiredness in my legs, more hope in my steps than sadness in my shoulders, more road in my heart than fear in my head.

Happy is he who transfers what he knows and learns what he teaches.

(Cora Coralina)

SUMMARY

SUMMARY

MACHADO, D.M.C. (2013) *Evaluation of Defect Identification Standards for the Management of Flexible Pavements.* 126 P. Dissertation (Master's Degree) - São Carlos School of Engineering, University of São Paulo.

This work analyses methods for assessing the condition of flexible asphalt pavements, looking at the various types of defects that can appear on them, their causes and the most suitable maintenance and rehabilitation activities, depending on their severity and extent. It is part of the 'Pavement Management' theme, and it is of great importance to know the condition of pavements in order to make decisions about the investments needed to guarantee vehicle traffic with safety, comfort and, above all, economy, i.e. with low vehicle operating costs. Evaluating pavements makes it possible to identify constructive faults that manifest themselves early on, as well as monitoring the natural wear and tear that all pavements undergo, helping to ensure that the inevitable interventions arc well selected and carried out at the right times, thus enabling the best possible use of available resources. This paper studies pavement evaluation methods used in Brazil, the United States and France, both for network and project-level pavement management, for road, urban and airport pavements, based on objective and subjective evaluations, considering structural and non-structural defects. A critical analysis was carried out with the aim of contributing to a simple and efficient method for assessing the condition of pavements in Brazil, developed with academic rigour but easy to use by the technical community. The negative points of the methods currently used in Brazil were identified, and alternatives to solve existing problems were sought in international standards. Finally, it is recommended that the Brazilian Standard be revised, with changes so that it can be applied in a homogeneous way by the different technicians who use it, making it more agile, simple and efficient. The proposed changes include: a clearer definition of cracks, grouping Corrugation and Slippage defects, separating defects of structural and non-structural origin, more effective application of severity levels, and pavement assessment using the ICP concept.

Keywords: Flexible Pavements, Management System, Evaluation, Defects.

CHAPTER 1

INTRODUCTION

1.1 Initial considerations

The concept of a good quality road encompasses not only comfort and safety, but also economic efficiency, which is becoming increasingly important in Brazil, given the majority use of road transport for the transport of goods and people.

Defects found in road surfaces therefore compromise safety and comfort and can cause major economic losses. Constant assessments are needed to show how much safety, comfort and operating costs are jeopardised by defects, how these defects can be repaired, the corresponding costs and where the priorities lie.

Within this analysis, it is worth mentioning the concept of Pavement Management, which involves the economic optimisation of road transport, determining the correct time for interventions on the pavement, defining what these interventions will be and how they will be carried out, in order to benefit all sectors of society.

It is possible to assess, both objectively and subjectively, the defects that pavements present during their useful life. However, in order to effectively assess these defects, they need to be consistently classified.

The research carried out for this study revealed the existence of different classifications of defects that appear in pavements and different forms of assessment to which these pavements are subjected. The multiplication of classifications and evaluations is not a problem in itself, but many of these approaches have theoretical flaws and conceptual inconsistencies. It is clear, therefore, that there is a need for a critical analysis of the defects in flexible pavements that are considered, their formal definition, and the method for identifying the type, severity and extent of each one specifically.

1.2 Objectives

The general aim of this work is to propose ways of better classifying defects in flexible pavements and ways of identifying these defects that can be used for both academic and technical purposes. At the same time, proposals are made for evaluating existing defects in flexible pavements in a clear and efficient manner. The intention is to achieve a standardised treatment that can be applied to any circumstances.

1.3 Method

In order to achieve the objective of proposing adjustments for a clearer classification and a simpler and more efficient assessment of defects that occur in asphalt pavements, these defects are described and the standards for their assessment adopted in the United States, France and Brazil are analysed.

Some inconsistencies are pointed out that could, in some cases, jeopardise pavement management in Brazil.

A case study is presented, with functional and structural assessments carried out on a state highway (DER/SP - Departamento de Estradas de Rodagem do Estado de São Paulo). The data obtained through routine assessments carried out by the public bodies that manage the road network is compared with the data obtained through the assessments suggested in this work.

At the end, suggestions are made so that Brazilian standards can better meet Brazilian needs in terms of Pavement Management.

Chapter 2 details the pavement defects that occur most frequently and most affect safety, comfort and economy when travelling on Brazil's highways. Some defects are defined differently by the various standards, which can jeopardise the efficiency of identifying them. Standardisation, in the most appropriate way possible, is an objective to be achieved in order to speed up the recovery processes of the Brazilian road network.

CHAPTER 2

DEFECTS IN FLEXIBLE PAVEMENTS

2.1 Initial considerations

In order to effectively deal with the rehabilitation of road surfaces, with the aim of maintaining their good traffic condition, it is necessary to correctly assess what is happening on a given road segment.

When it comes to identifying defects in road pavements, it is necessary to define which ones may arise. Each defect needs to be identified, along with its extent and level of severity. There are various classifications, both national and international, which often makes it difficult to deal with the subject. When analysing Brazilian standards, in this case the DNIT Manual and the agency's Standards that define defects and detail how they should be assessed, it is possible to see inconsistencies regarding some defects, making it difficult to analyse and deal with them properly.

In this work, comparisons were made between the various approaches to the subject taken by US, French and Brazilian technical bodies.

Initially, a study was made of the approach taken by the FHWA (Federal Highway Administration) within the LTPP (Long Term Pavement Performance) programme of the SHRP (Strategic Highway Research Program), which was developed in the USA specifically to address these difficulties in dealing with the management of defects that arise in road pavements. It was possible to see that the Defect Identification Manual used by the SHRP contains a fairly objective classification of these defects.

Another analysis was also carried out, based on French standards, which differentiate between defects 'caused by structural problems' and those 'caused by problems other than structural'. This division could be important when deciding 'what type of rehabilitation should be carried out on a motorway', as it better identifies 'the cause of the defect'.

Finally, a critical analysis was made of the treatment given to the subject by the Brazilian standards regulated by the DNIT. The purpose of this analysis was to propose some changes in order to speed up the implementation of Pavement Management in Brazil.

2.2 Defects in Asphalt Concrete Pavements - US Standards

According to the manual drawn up by the FHWA in the USA, the 'Distress Identification Manual' for the LTPP within the SHRP, defects are subdivided into Cracks, Patches and Potholes (or Pans), Surface Deformation and Corrugation (or Slippage), Surface Defects and Other Defects. Following on from this work, each defect is detailed in accordance with the manual, with the addition of some considerations from the work of the University of São Paulo, from 2003, duly cited in the attached bibliography, called "Defects and Maintenance and Rehabilitation Activities on Asphalt Pavements". Some indications on how to size defects and how to classify their severity levels have also been introduced according to one of the attached bibliographies, "Modem Pavement Management", which also deals with Pavement Management.

2.2.1 - Introduction

Initially, the SHRP programme, which began in 1987, was planned to work for five years. However, the objectives to be achieved for the LTPP led to an additional fifteen years. In 2007, it was extended for a further twenty years.

When dealing with each defect that occurs on the pavement, it is essential to define the level of severity and the size of the defect: extent in metres or square metres, or quantity. It is also detailed how each defect is to be repaired.

The list of pavement defects used in the SHRP (FHWA, 1993), as well as its translation and adaptation for Brazil (FERNANDES JR. et al., 2003), are shown in Table 1.

Table 1 - Defects in Flexible Pavements

> Cracking
1. Fatigue cracking
2. Block Cracking
3. Edge cracking
4a. Wheel Path Longitudinal Cracking

4b. Non-Wheel Path Longitudinal Cracking

5. Reflection Cracking at Joints
6. Transverse Cracking
> Patching and Potholes
7. Patch / Patch Deterioration
8. Potholes
> Surface Deformation (permanent surface deformation)
9. Rutting (permanent deformation)
10. Shoving (corrugation)
> Surface Defects
11. Bleeding
12. Polished Aggregate
13. Ravelling (wear)
> Miscellaneous Distresses
14. Lane-to-Shoulder Dropoff
15. Water Bleeding and Pumping

Each of the defects is detailed below.

2.2.2 - Cracks and subdivisions

This defect is associated with openings that appear in the asphalt surface. Some bibliographies refer to 'fissures' when the opening is only perceptible to the naked eye at a distance of less than l.5m and 'cracks' when the opening is greater than the fissure (BERNUCCI et al., 2007 and DNIT-005/2003-TER, 2003).

Cracks are classified as: Fatigue cracks, Block cracks, Edge cracks, Longitudinal cracks, Joint reflection cracks, Transverse cracks.

These types of cracks are described below according to the FHWA (1993) and Femandes Jr. et al. (2003).

a) *Fatigue cracks* appear as small, irregular, connected blocks, similar in appearance to alligator or crocodile skin, spaced less than 0.3 metres apart. This type of defect usually occurs when the pavement, subjected to repeated or continuous application of loads due to road traffic, reaches the limit of its capacity. It can also occur due to other factors: the loads to which the pavement was subjected were above its design structural capacity, the pavement layers were dimensioned with thicknesses less than those required for the existing traffic, the construction was not accompanied by adequate quality control, the water (rain) weakened the pavement structure. Cracks are usually found in the alignment of vehicle tyres (especially heavy vehicles), a place called the 'wheel track'. Figure 1, taken from Standard DNIT 005/2003-TER, illustrates this defect.

Figure 1 - Fatigue cracks

Source: Standard DNIT 005/2003 - TER (2003)

The severity of this defect can be: low (a few cracks, disconnected or slightly connected, no pumping and no erosion at the edges), moderate (there are a significant number of connections, forming a more defined pattern, there may be some wear and sealing of the connections, and also some erosion at the edges, there is no pumping), high (the crack pattern is evident, there is sealing and accentuated wear, there are also accentuated erosions at the edges and pieces are removed with the action of traffic, there is pumping).

Fatigue cracks should be measured by checking the area of occurrence of each severity level (low, moderate or high). They are repaired with surface treatment or asphalt slurry (temporary), or patching (permanent). In some cases, resurfacing or total reconstruction of the pavement may be necessary.

b) *Block cracks* are arranged in the approximate shape of connected rectangles (0.1 to 10 square metres in area). This type of defect is often caused by contractions in the asphalt mixture beyond what it can withstand. They can be of thermal origin in the surface layers, where the surfacing may have been formed from asphalt mixtures with fine aggregates and asphalt with low penetration, or they can be caused by variations in the moisture content in the lower layers.

Loss of elasticity, which results in block cracking, can be caused by: too long a mixing time, mixes made at temperatures above those recommended by the paving standards, storage above the time indicated by the paving standards. The base may have been treated with cement or used tropical soils, which can also cause shrinkage. Figure 2, taken from Standard DNIT 005/2003-TER, illustrates this type of defect.

Figure 2 - Cracks in blocks Source: Standard DNIT 005/2003 - TER (2003)

The severity of this defect can be: low (cracks measuring up to 6 mm or sealed cracks with undefined dimensions), moderate (cracks measuring between 6 and 19 mm or a few cracks measuring less than 19 mm added to a significant number of cracks with low severity), high (cracks measuring more than 19 mm or a significant number of cracks measuring less than 19 mm added to a significant number of cracks with moderate to high severity). Block cracks should be measured by checking the area of occurrence of each severity level (low, moderate or high) and the standard area of the blocks in each section.

In areas with small cracks and deep cracks, sealing material can be used (asphalt emulsion, surface treatment, asphalt slurry). In more severe areas, patching, recycling, resurfacing or even reconstruction of the pavement may be necessary.

Edge cracks only occur when shoulders are not paved. They appear at the edge of the pavement (at the boundary with the shoulder), in irregular and increasing or regular sizes, in a strip 0.60 metres from the edge of the pavement. They appear as a result of insufficient compaction or poor drainage. Figure 3, from the DER/SP archive, illustrates this type of defect on SP 338.

Figure 3 - Edge cracks

Source: DER/SP archive

Severity can be: low (cracks without ruptures or loss of material), moderate (with some ruptures and loss of material over up to 10 per cent of the length of the affected pavement), high (cracks with ruptures and considerable loss of material over more than 10 per cent of the length of the affected pavement). Edge cracks should be measured by checking the length (metres) of the affected pavement separately for each severity level.

Repair with sealant to prevent water ingress and structural weakening. In the event of poor drainage, this must be corrected. In some cases it is necessary to patch the damaged area.

c) *Longitudinal cracks* appear predominantly parallel to the axis of the road. Their location is well defined: inside or outside the wheel tracks. These cracks can facilitate the penetration of water through the surface layer of the pavement, causing instability between the various layers of the pavement, and accelerating the development of fatigue cracks or the appearance of surface wear. In the case of longitudinal cracks between adjacent lanes, outside the wheel tracks, the probable cause of these cracks is inadequate compaction when the pavement was laid, or compaction carried out in a very different way between each lane. In the case of longitudinal cracks at the edge of the wheel tracks, the probable cause is excessive load applied, especially by heavier vehicles. It can also be aggravated by mixtures with very high temperatures applied to a very fragile base. Figure 4, taken from Standard DNIT 005/2003- TER, illustrates longitudinal cracking outside the wheel tracks.

Figure 4 - Longitudinal crack outside the wheel tracks
Source: Standard DNIT 005/2003 - TER (2003)

The severity of this defect can be: low (cracks measuring up to 6 mm or sealed cracks with undefined dimensions), moderate (cracks measuring between 6 and 19 mm or a few cracks measuring less than 19 mm added to a significant number of cracks with low severity), high (cracks measuring more than 19 mm or a significant number of cracks measuring less than 19 mm added to a significant number of cracks with moderate to high severity). Longitudinal cracks, both inside and outside the wheel track, should be measured by quantifying their length (metres) and separating them by severity level. The length of sealed cracks should also be measured.

Repairs should be carried out as follows: cracks smaller than 3 mm do not need to be repaired; cracks between 3 and 20 mm should be sealed; when the cracks are larger than 20 mm, the pavement should be patched, resurfaced or rebuilt.

d) *Joint Reflection Cracks* are caused by discontinuities in the lower layers, which propagate through the asphalt coating. This defect can occur as a result of cracks or joints in the lower layers of rigid pavement, if the asphalt pavement was built on concrete pavement, or bases with greater rigidity, such as bases treated with cement or lime.

The severity of this defect can be: low (cracks measuring up to 6 mm or sealed cracks with undefined dimensions), moderate (cracks measuring between 6 and 19 mm or a few cracks measuring less than 19 mm added to a significant number of cracks with low severity), high (cracks measuring more than

19 mm or a significant number of cracks measuring less than 19 mm added to a significant number of cracks with moderate to high severity). Figure 5, taken from FHWA (1993), illustrates reflection cracks in joints.

Figure 5 - Reflection cracks in joints

Source: FHWA (1993)

Reflection cracks in joints are measured by considering the type of crack that caused them. They can be assessed in the same way as longitudinal or transverse cracks, separately, and measured by level of severity. In the case of longitudinal cracks, the extent (inside or outside the wheel tracks) is recorded; in the case of transverse cracks, the extent and quantity are recorded. The extent of sealed cracks should also be recorded. Reflection cracks are also recorded as block cracks or fatigue cracks, if applicable.

To repair reflection cracks in joints, patching and surface treatment or asphalt slurry should be used. In order to prevent defects from the old pavement from spreading to the new pavement (reflection), appropriate techniques should be used (geomembranes, recycling, others). In the case of an occurrence that has already occurred, patching, resurfacing or reconstruction of the pavement should be used, depending on the type of crack that has 'reflected' on the upper pavement.

f) *Transverse cracks* appear predominantly perpendicular to the axis of the road. They can occur due to thermal contraction of the surfacing and water infiltration into the lower layers.

The severity of this defect can be: low (cracks measuring up to 6 mm or sealed cracks with undefined dimensions), moderate (cracks measuring between 6 and 19 mm or a few cracks measuring less than

19 mm added to a significant number of cracks with low severity), high (cracks measuring more than 19 mm or a significant number of cracks measuring less than 19 mm added to a significant number of cracks with moderate to high severity). Figure 6, taken from Standard DNIT 005/2003-TER, illustrates this type of defect.

Figure 6 - Transverse crack Source: Standard DNIT 005/2003 - TER (2003)

Transverse cracks are measured by quantifying the number of occurrences and their length (metres) for each severity level. There must be more than 10 per cent of the severity indicated to be classified as such. The length of the sealed cracks at each severity level must also be indicated. Transverse cracks less than 30 cm long are not recorded.

Repairs to cracks that have already been installed are carried out using sealant material at points of reduced size and crack depths. At more severe points, patching, recycling, resurfacing or even reconstruction of the pavement may be necessary. In the event of water seeping into the lower layers, drainage must be corrected to prevent the defect from recurring.

2.2.3 - Need patching (or damage that can be patched), Potholes

In the Defects Manual used in the USA (FHWA, 1993), patches and potholes are grouped together as similar defects. A patch is a type of defect, although it is related to surface maintenance, and is characterised by the filling in of potholes or any other hole or depression with asphalt mix. A pothole is a cavity that appears in the asphalt surface and may or may not reach the underlying layers (BERNUCCI et al., 2007). The two types of defects are detailed below, according to FHWA (1993) and Femandes Jr. et al. (2003).

a) *Patching* is a portion of surface pavement, larger than O.lm^2, that is removed and replaced with material applied over the original construction of the road. Various defects can cause the *need for patching.*

Severity can be: low (dimensions of less than 6 mm have little effect on traffic quality), moderate (dimensions of between 6 mm and 12 mm have a moderate effect on traffic quality), high (dimensions of more than 12 mm and occasional pumping also have a serious effect on traffic quality). These limits are indicated in the FHWA (1993). Severity can also be measured according to the defects to be corrected. Pumping is not evident in cases of low and moderate severity. To quantify this defect, a record is kept of occurrences and the affected area for each level of severity.

The repair is carried out by cutting out and cleaning the affected area, applying the waterproofing and adhesive layer and then the replacement material. When the material is applied without cutting out the affected area, only after cleaning, the name given to the repair is 'pothole cover'. Figure 7, taken from Bemucci et al. (2007), illustrates a 'well-executed patch'.

Figure 7 - 'Well' executed patch

Source: Bemucci et al (2007)

b) *Potholes* are the result of localised disintegration under the action of traffic and in the presence of water. They show fragmentation of the pavement after fatigue or wear cracks and localised removal of parts of the coating. They occur in bowl shapes of various sizes on the pavement surface. They most commonly occur in coatings with a low thickness or low bearing capacity of the lower layers (structural failure). They can appear in places with material segregation (lack of binder at some points) or with construction problems (inadequate drainage). Figure 8, taken from Standard DNIT 005/2003-TER, illustrates this type of defect.

Figure 8 - Hole or Pan

Source: Standard DNIT 005/2003 - TER (2003)

The level of severity can be: low (up to 25 mm deep), moderate (25 to 50 mm deep), high (over 50 mm deep). These limits are indicated in the FHWA (1993) and are consistent with the way in which this type of defect is treated in the USA; however, for the Brazilian reality, these values may be altered for longer intervals.

To measure this defect, the number of occurrences and the affected area for each level of severity should be recorded. Repairs should be carried out by patching, followed by resurfacing if necessary, depending on the size of the affected area.

2.2.4 - Permanent Deformation, Corrugation (or slip)

As with the previous item, the US Defects Manual (FHWA, 1993) groups these occurrences as similar defects. These defects cause depressions in the pavement and soiling of the asphalt mix (BERNUCCI et al., 2007). These two types of defects are detailed below, according to the bibliographies of the FHWA (1993) and Femandes Jr. et al. (2003).

Permanent Deformation appears in the path of the wheels, or wheel track. It is a longitudinal subsidence that follows the path travelled by vehicle wheels. It usually occurs due to densification of materials or shear rupture. It can result from: inadequate sizing of the thicknesses of the various layers, inadequate dosage of the asphalt mix, inadequate compaction and subsequent consolidation by traffic, drainage failure. In the case of shear failure, subsidence in the wheel tracks is accompanied by uplift on the sides, parallel to the traffic. Figure 9, taken from Standard DNIT 005/2003- TER, illustrates permanent deformation in the wheel tracks.

Figure 9 - Permanent deformation in wheel tracks
Source: Standard DNIT 005/2003 - TER (2003)

Severity levels can be: low (up to 25 mm deep), moderate (between 25 and 50 mm deep), high (over 50 mm deep). Measurements of the gradient or maximum depths of deformation, in millimetres, should be recorded at 20 m (twenty metre) intervals. These limits are indicated in Haas et al. (1994). The repair should be carried out by recycling, resurfacing or completely rebuilding the pavement.

b) *Corrugation or Slippage* is a longitudinal displacement of the pavement, causing transverse undulations on its surface. This displacement is usually caused by the 'braking' or 'acceleration' movement of vehicles and is due to inadequate dosage of the asphalt mix, poor bonding between base and surfacing, and insufficient structure. It is usually located on curves, intersections, slopes or inclines.

Severity levels can be determined according to how much traffic quality is affected. Figure 10, taken from the DER/SP archive, illustrates the 'corrugation' defect with the appearance of some 'potholes' and drainage problems.

Figure 10 - Corrugation on a motorway access device

Source: DER/SP archive

This defect is measured by recording the number of occurrences and the area affected (in square metres). The affected sections must be patched, recycled, resurfaced or completely rebuilt.

There is a classification in US agencies that is similar to corrugation. This is "slippage cracking", which occurs in areas where vehicles accelerate: the vehicle goes forward and the friction force 'pushes' the pavement backwards (it slips).

2.2.5 - Surface defects: Exudation, Polished Aggregates, Wear

Again, the US Defects Manual (FHWA, 1993) groups similar defects together. Surface Defects include: exposure of excess binder, exposure of aggregates or detachment of aggregates (BERNUCCI et al., 2007). These three types of defects are detailed below, according to FHWA (1993) and Femandes Jr. et al. (2003).

a) *Exudation* occurs when there is excess bituminous material on the pavement surface, usually in the wheel track. It can appear as asphalt that has changed its normal colour or lost its normal texture. Often the aggregate shows this excess bituminous material in its darker colour. The pavement surface is shinier and sometimes sticky. Tyre marks can appear at higher temperatures and can lead to a reduction in tyre-pavement friction. This defect is usually the result of an inadequate asphalt mix (low void ratio).

Severity levels can be classified according to the local situation or the evolution of the problem, as follows: 'low level' is when there is a change in the colour of certain sections of the pavement compared to the general situation, 'moderate level' is when there is a change in the texture of the pavement, 'high level' is when the pavement has a shiny appearance and tyre marks appear on it (especially on days with high temperatures). Figure 11 shows three situations in which exudation occurs, taken from DNIT 005/2003-TER (2003) and Bemucci et al. (2007).

Figure 11 - Occurrences of Exudation

Sources: Standard DNIT 005/2003 - TER (2003) and Bemucci et al. (2007)

To quantify this type of defect, the area of occurrence is checked, classified by level of severity. Exudation must be repaired with surface treatment or hot sand, or even recycling.

b) *Polishing or wearing of the aggregate and bituminous binder* can expose the coarse aggregate, often resulting in loss of material. This defect can be the result of the abrasive action of traffic, which removes the roughness and angularity of the particles. Figure 12 illustrates the occurrence of this type of defect and was taken from the FHWA (1993).

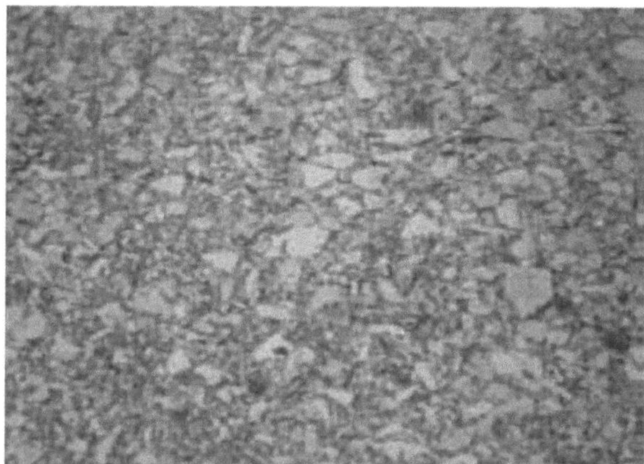

Figure 12 - Polished aggregates

Source: FHWA (1993)

There is no way of defining severity levels. The degree of polishing can lead to a reduction in tyre-pavement friction, compromising road safety. The areas affected, in square metres, are assessed.

To repair this defect, surface treatment or asphalt slurry should be carried out. In some cases it is advisable to recycle, resurface or reconstruct the pavement.

c) *Wear* is caused by the displacement of aggregate particles and loss of adhesion of the asphalt binder. Wear can evolve from the loss of fines to the loss of some coarse aggregate, and finally to more significant aggregate losses. This is due to ageing, hardening, oxidation, volatilisation and weathering. The more suitable the asphalt mixture (dosage or temperature) and its subsequent application (good weather conditions at the time of application, correct compaction, adequate time for release to traffic), the better the adhesion between aggregates and asphalt binder, which reduces the likelihood of early wear. Figure 13, taken from Standard DNIT 005/2003-TER, illustrates the appearance of wear.

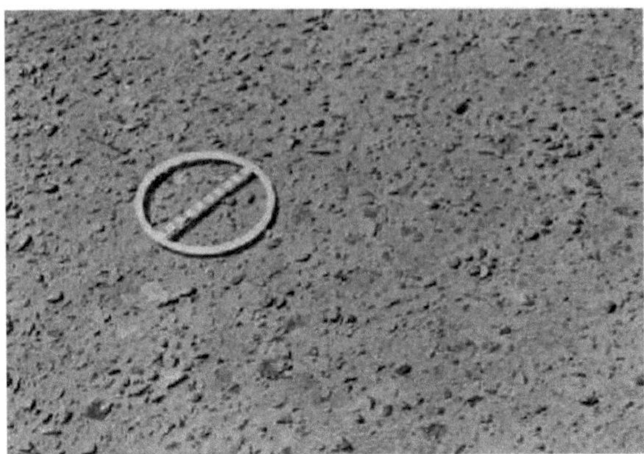

Figure 13 - Wear

Source: Standard DNIT 005/2003 - TER (2003)

Severity levels can be classified according to the local situation or the evolution of the problem, as follows: 'low level' is when there is a loss of fine aggregates, 'moderate level' is when the surface texture becomes rougher and there is a loss of some coarse aggregates, 'high level ' is when the surface texture becomes very rough and there is a more intense loss of coarse aggregates. The measurement is made by quantifying the area of occurrence, classified at each level of severity. Repairs can include: sealcoating, surface treatment or asphalt slurry, recycling or resurfacing.

Within the concept of 'wear', there is a subdivision in some North American agencies that classify two types of this defect. One is "stripping", which is a type of wear that occurs when there is incompatibility of materials and, under the action of water, wear begins (occurs early). The other is "wear loss", which is the dislodgement of aggregates due to ageing and loss of adhesion (this wear occurs as the pavement ages).

2.2.6 - Other defects: Gap between carriageway and hard shoulder, Pumping

In a final segment, the US Defects Manual (FHWA, 1993) groups together two other defects not covered by the previous situations. Their descriptions, again, are from the FHWA (1993) and Femandes Jr. et al. (2003) bibliographies.

a) The *gradient (or step) between the carriageway* and the shoulder is the difference in elevation between the carriageway and the shoulder. Normally, this unevenness results from the application of different layers of material, and sometimes a greater thickness is applied to the road surface than to the shoulder. In the case of an unpaved shoulder, there may be erosion on the shoulder, causing the gap between the shoulder and the carriageway. There may also be differential consolidation between the carriageway and the shoulder. Figure 14, taken from the DER/SP archive, illustrates this type of defect.

Figure 14 - Gap between carriageway and unpaved shoulder

Source: DER/SP archive

Severity levels are not applied in this case. They are replaced by measurements of the heights of the gradients every 20m (metres). Repairing this defect consists of recomposing the shoulder.

b) *Pumping* is related to the ejection of water from the underside of the pavement through cracks when under the action of traffic loads. In some cases, this problem can be detected by the presence of fine material on the pavement surface that has been eroded from the supporting or base layers. This defect occurs when there is water in the voids under the surfacing, i.e. there is a deficiency in the drainage system.

It is difficult to define severity levels because the amount of water pumped can vary according to humidity conditions. The FHWA (1993) provides an illustration of this defect in Figure 15.

Figura 15 - Pumping

Source: FHWA (1993)

To quantify this defect, the number of occurrences and the area affected are recorded. The pavement should be repaired as necessary and the drainage system should be corrected to avoid pumping occurring again.

2.3 Defects in Asphalt Concrete Pavements - French Standards

In 1972, the LCPC (Laboratoire Central des Ponts et Chaussées) published a 'catalogue of pavement degradation', drawn up by the CCT (Centre Coordonnateur des Trappes) with the help of engineers from the 'Reinforcement Support Points' (PAR) of the Regional Laboratories. This catalogue was followed in 1977 by the 'Flexible Pavement Survey Guide', in which a chapter was dedicated to the classification, survey and quantification of pavement surface defects. After the era of reinforcements came the era of preventive maintenance of national roads and, in 1979, the Directorate of Roads and Road Traffic published a collection of degradations, prepared and distributed by the LCPC and SETRA (Service D'Estudes Techniques des Routes et Autoroutes), the latter being the body that deals with the Directorate of Roads. The 1998 version of this collection (or catalogue) of degradations was used. This was followed by VIZIR, a 'network quality assessment' method for subsequent decision-making.

2.3.1 - Introduction

French standards treat defects found in road surfaces differently . These defects are separated into:

structural degradations and degradations not linked to structural problems. In the specifications described below, some types of defect do not have their causes detailed.

2.3.2 - Structural damage

According to LCPC (1998) and Autret & Brousse (2001), *structural degradation* can be localised only on the surface (surfacing) or affect the deeper layers (base, sub-base, subgrade, subgrade reinforcement). They occur when the pavement's structural capacity is insufficient. These types of defects are subdivided into: fatigue cracks, permanent deformation and patches. They usually occur in wheel tracks, especially when heavy vehicles are travelling. These defects are illustrated in Figures 16, 17 and 18. The illustrations were taken from Bemucci et al. (2007).

Figura 16 - Fatigue cracks - second illustration

Source: Bemucci et al. (2007)

Figura 17 - Permanent deformation - second illustration

Source: Bemucci et al. (2007)

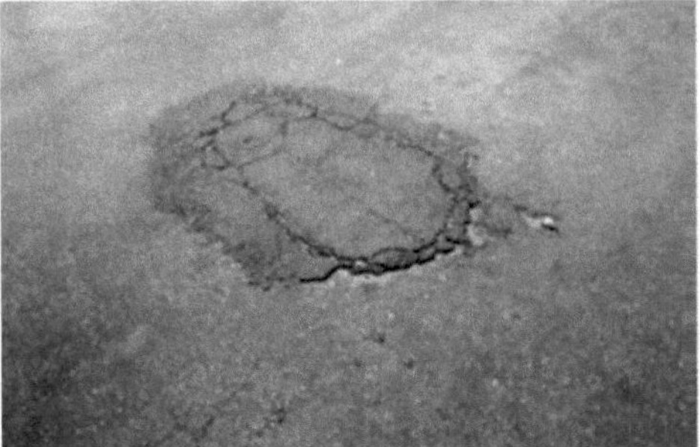

Figura 18 - Poorly' executed patch

Source: Bemucci et al. (2007)

The three types of defect are described below according to LCPC (1998) and Autret & Brousse (2001).

a) *Permanent deformation* occurs as a result of insufficient support of the layers below the surfacing, poor quality of the support layers and reduced bearing capacity due to poor drainage. The causes can vary depending on the type of base. This deformation appears in the wheel track, at a distance of 50 to 80 cm from the edge. It arises due to the settlement of materials under heavy, channelled traffic, whether or not associated with a drop in the resistance of the lower layers.

Severity varies from low (up to 2 cm in height), medium (2 to 4 cm in height) and high (over 4 cm). At this point it's worth noting that, for SHRP, the severity levels can be: low (up to 2.5 cm), moderate (between 2.5 and 5.0 cm), high (over 5.0 cm), i.e. quite similar.

Within this classification, we can include 'localised or punctual sinking' ("affaissement hors rive" or "flache"), which can also originate from the same causes as above with the difference that it is restricted to a small area (almost a point). It may or may not be localised in the wheel track.

b) *Fatigue cracks* occur due to the constant action of road traffic, which subjects the pavement to repeated (continuous) loads. Over time, cracks can form on the pavement surface, giving rise to 'fatigue cracks'. They can develop into an 'alligator or crocodile skin' appearance. They preferably appear in wheel tracks.

Within this classification, there is a type of circular crack ("faiençage circulaire") that ends up leading to fatigue cracking occurring more widely, usually also in the wheel track. Figure 19, taken from the LCPC (1998), schematises this type of circular crack.

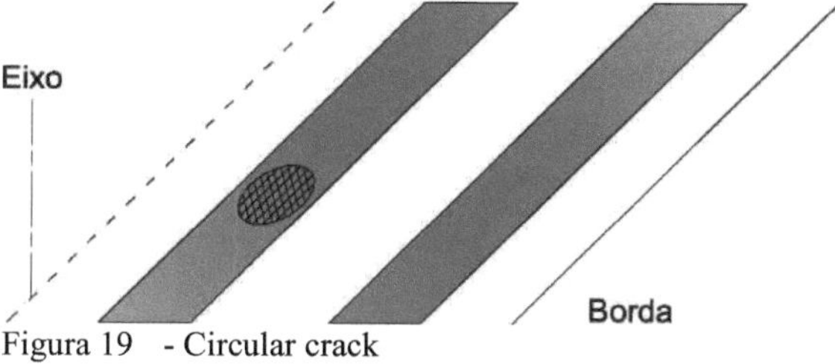

Figura 19 - Circular crack

Source: LCPC (1998)

c) *Patches* are made to regularise the pavement when there are surface or deep defects. A patch can 'mask' a problem. If patches are frequent in the same place, it may mean that the problem has not been solved. Constant patching can indicate the need for repairs to the pavement structure.

2.3.3 - Degradation not linked to structural problems

According to LCPC (1998) and Autret & Brousse (2001), *degradations not linked to structural problems* are defects not associated with problems with the pavement's structural capacity. They are caused by problems at the time of construction, whether it be: the quality of a particular material used, the way the asphalt mix was prepared or the way the base and surfacing were made. They can also arise as a result of specific local conditions and traffic. They can include cracks other than 'fatigue cracks' (longitudinal, block, transverse, edge), wear, exudation, corrugation, peeling, combing, potholes, unevenness between carriageway and shoulder.

The various defects classified as *not originating from structural problems* are explained below from the point of view of the LCPC (1998) and Autret & Brousse (2001).

a) *Longitudinal cracks* are described as a type of defect resulting from an imperfection in the bonding of two contiguous premixed strips. In this case, the longitudinal crack does not necessarily appear in the wheel track. In cases where the longitudinal crack appears in the wheel track, it is likely that there is a defect of the 'structural degradation' type.

b) *Edge cracks* appear close to the shoulder when it is not paved. They can be caused by inadequate sizing of the pavement layers, poor compaction, drainage problems, specific climatic situations or excessive vegetation on the edges making it difficult for water to escape.

c) *Potholes* appear as a result of wear or destruction of the bearing layer. Initially, their size is small. If not corrected, their size can increase considerably, even leading to the removal of the pavement. It starts in the surfacing, but can reach the base if not repaired.

It's important to emphasise that although in most cases the *pothole* defect is caused by various problems, in some cases it can be *structural.*

d) *Transverse cracks* can occur due to very large temperature variations, especially in countries with cold climates.

e) *Block cracks* comprise a set of longitudinal and transverse cracks that join together to form a mesh network. They arise due to shrinkage after the asphalt mixture has been applied.

f) *Wear* initially appears as a small loss of material. Over time, this loss accelerates and jeopardises the quality of the pavement.

g) *Exudation* occurs when excess binder appears on the surface.

h) *Corrugation* can appear as a result of poor compaction, when the pavement is under heavy pressure, especially on sharp bends or ramps, or in places subject to constant braking and acceleration.

Within this classification, some specific defects ("gonfle") can be included, not always located in places subject to constant braking and acceleration, but with similar characteristics (it looks like a "puff" in the pavement). They arise due to chemical incompatibilities or even poor compaction.

i) *Peeling* occurs when parts of the road surface loosen (peel off) from the base or previous coating, in the case of preventive maintenance or thin resurfacing (asphalt micro-coatings and surface treatments, for example). This defect is considered a pothole in North American standards, even though it is of low severity due to its small thickness. It can also occur when some 'pans' of low severity are repaired and the material applied does not adhere properly to the lower layers. It is therefore suggested that, in Brazilian standards, this defect be included alongside the Hole or Pan defect.

j) *Combing* occurs due to poor execution of the asphalt layer in surface treatment, usually caused by irregular distribution of the binder by the spreader (failure of the spreader nozzle), or due to inadequate distribution of aggregates in the pavement. This defect is not considered in the US standards, which only consider hot-machined asphalt concretes and do not include surface treatments. Figure 20, taken from Loiola (2009), illustrates this defect.

Figure 20 - Combing
Source: Loiola (2009)

k) *Gap between carriageway* and shoulder (more common in the case of a paved shoulder) and *erosion on the shoulder* (in the case of an unpaved shoulder) usually arise due to construction faults or deficiencies in routine maintenance. Both defects can occur at the boundary between the carriageway and the shoulder. In the case of a paved shoulder, there can be a steep gradient between the two, which causes a dangerous step for the road user to cross. In the case of an unpaved shoulder, there may be erosion on the shoulder, which will also compromise traffic safety, and this erosion could reach the road, further aggravating the situation. There may also be differential consolidation between the carriageway and the shoulder, according to FHWA (1993) and Femandes Jr. et al. (2003).

The severity of this defect should be mentioned, as it is not mentioned in other standards. The gradings are: severity 1 or low severity with unevenness between 1 and 5 cm, severity 2 or medium severity with unevenness between 5 and 10 cm, severity 3 or high severity with unevenness above 10 cm.

2.3.4 - Comparison of severity levels between France and other countries

Within the specifications of the French technical bodies, there are some guidelines on how to measure the various types of defects (severity, extent, others). These guidelines are not detailed in this dissertation, as the US guidelines are simpler and more precise. In the case of fatigue cracks (structural degradation) or other cracks (edge cracks, block cracks, transverse cracks, longitudinal cracks, reflection cracks), the French standards do not indicate the values (thickness or width) to be taken into account when classifying severity. These standards use the expressions 'clearly open' or 'very open' to measure fatigue cracks, which can leave room for varying interpretations between assessors. The US standards, on the other hand, use parameters to classify these severity levels, leaving fewer details to be assessed by the 'common sense' of the technicians responsible.

The dimensions for classifying potholes into severity levels are only clearly defined in the US manuals. In the case of corrugation, the French standards do not indicate any values, nor do the US standards, but the latter link severity to 'how much safety is affected'. In the case of exudation and wear defects, although no values are given in any of the standards analysed up to this point, in the case of the North American standards these defects are constantly monitored. The severity levels of these defects are indicated in Femandes Jr. et al. (2003). Polished aggregates and pumping are only mentioned in the North American standards, but it is important to assess these defects in view of their constant occurrence in Brazilian pavements. In the case of polished aggregates, the variation in tyre-pavement friction is assessed, while in the case of pumping there is no way of assessing severity. Peeling, which is only mentioned here, is classed as a hole or pothole for comparison with other standards. Combing is also only mentioned in the French standard and no values are quantified to classify its severity levels.

The classification of the severity of permanent deformation, between 2 and 4 cm, is very similar to that adopted by the United States, between 2.5 and 5 cm (HAAS, 1994), for comparison purposes. The Pavement-Sidewalk Gap defect cites values for classification by severity level only in the French standards. In this specific case, the French guidelines, which are quite simple, should be used.

A final consideration is the analysis of patches. The French Standard states that patches can be classified into two types: one that completely suppresses the defect is probably a patch referring to degradation of the 'non-structural origin' type, and another that partially suppresses the defect is probably a patch referring to degradation of the 'structural origin' type. This differentiation can be important when making decisions about pavement rehabilitation and should be taken into account when classifying by level of severity. Numerical values are only given in the US standards.

2.4 Defects in Asphalt Concrete Pavements - DNIT / Brazil Standards (DNIT, 2006; NORMA DNIT-005, 2003)

One of the aims of the DNIT's Road Pavement Restoration Manual, used as a source for this topic, is to present and discuss the technical elements needed to identify, quantify and analyse existing deterioration in asphalt pavements. Techniques for assessing the structural capacity of pavements, methods for determining their ability to provide rolling comfort and safety, methods for sizing pavement reinforcement and quality control measures for restoration services are discussed. Another objective is to present a process for selecting the best restoration strategy for flexible pavements. Within the development of the Manual, the main point noted in this work is the presentation of the main defects that can occur in asphalt pavements during their useful life, their causes, structural and functional pavement assessment processes.

2.4.1 - Introduction

In order to carry out a good pavement repair, it is necessary to identify the cause of the specific deterioration that has occurred on the site. It is also necessary to determine when exactly to intervene and what intervention will be carried out. It is worth highlighting the concept of Pavement Management, which works on optimising the use of resources and the exact timing of interventions to obtain the best economic return. An interrelationship between pavement performance, maintenance and rehabilitation strategy, intervention date and costs can be seen in Figure 21, taken from Femandes Jr. et al. (2003).

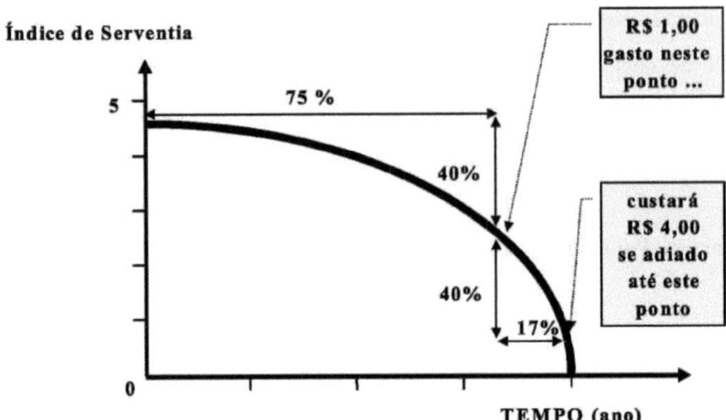

Figure 21 - Interrelationship between pavement performance, maintenance and rehabilitation strategy, date of intervention and costs

Source: Femandes Jr. et al. (2003)

2.4.2 - Classification

DNIT (2006) takes the following defects into account when carrying out restoration work:

- V Cracking (mainly due to fatigue)
- V Wear and tear
- V Hole (Pan)
- V Patches
- V Permanent deformation (sinking in the wheel tracks)
- V Functional irregularity ('associated with' undulation or corrugation and slippage)
- V Skid resistance ('associated with' exudation)

The first four are called surface defects (although they can also cover the lower layers). Permanent deformation and functional irregularity involve deformations in the lower layers. Skid resistance is compromised in the normal course of traffic and is a function of problems in the pavement texture.

It is important to emphasise that this classification is inconsistent in relation to the 'defects' that are actually meant. Functional Irregularity can be caused by Waviness or Corrugation and Slippage, which are 'defects' and not the irregularity itself. The same can be said for Skid Resistance, which is understood to be associated with the 'defect' Exudation (which will reduce this resistance and make the road unsafe). It is also worth adding that Skid Resistance, in many cases, is a surface defect without reaching the lower layers. In view of these comments, the defects that actually occur, and not those mentioned above, have been detailed.

Another classification is also cited by the DNIT, in an attempt to standardise with foreign standards, and is subdivided into two large blocks:

- J Cracks, which are subdivided into: *Isolated cracks* (transverse or longitudinal) and *Interconnected cracks* (fatigue or 'alligator skin' type, or block cracks);
- J Other defects, which are subdivided into: *Sinking* (plastic or consolidation), *Waviness or corrugation, Slipping, Exudation, Wear, Pans or holes, Patches.*

In the comments below, an attempt has been made to reconcile the two classifications presented by the DNIT, seeking the best coherence between them and the other international standards dealt with in this work.

2.4.3 - Cracking (DNIT, 2006; NORMA DNIT-005, 2003)

Cracking can take the form of cracks, fissures, isolated longitudinal or transverse cracks, interconnected cracks (resembling alligator leather) or even block cracks (DNIT, 2006).

A fissure is a capillary crack in the pavement that is perceptible from a distance of less than 1.50m. As it does not cause functional problems to the surface, it is not considered in terms of severity in pavement condition assessment methods (NORMA DNIT-005, 2003). A crack is a crack that is perceptible from a distance of more than 1.50m and can take the form of an isolated crack or an interconnected crack. (NORMA DNIT-005, 2003)

They can be classified as FC-1, FC-2, FC-3, depending on their dimensions (the number 1-2-3 takes into account the increased severity of the defect).

This type of defect can be caused by the following:

- J active traffic which, through the loading and unloading cycle, promotes tensile stresses in the inner fibre of the cladding;

J alternating daily temperature changes, which cause the existing coating to contract;

J reflection in the lining of existing cracks in cemented bases (soil cement base).

The DNIT Manual (2006) does not detail cracks, fissures or the various types of crack. The various types of cracks are detailed in Standard DNIT-005/2003-TER and are mentioned in the following items.

The DNIT Manual also mentions three types of crack: fatigue, ageing and reflection. The first type mentioned (fatigue) is related to the loading and relief cycle situation. In the second type of crack (ageing), the hardening of the asphalt that occurs with the loss of flexibility of the binder plays a greater role. The 'age' of this occurrence depends on: the oxidation resistance of the binder (chemical composition or origin of the oil), the ambient temperature and the film thickness of the binder. The third type of cracking (by reflection) occurs when there is cracking in the lower layers and this problem propagates into the upper layers.

This classification may cast doubt on the way cracking and its various subdivisions are classified, even changing established concepts about the causes and presentations of these various defects. Fatigue cracking arises as a result of loading and relieving, but it should not appear prematurely as the first item seems to suggest. This usually occurs if the pavement has been poorly dimensioned or if the loads to which the pavement has been subjected exceed what was initially designed. Over time, fatigue cracking tends to appear, due to the natural ageing of the pavement and the end of its useful life already foreseen in the project (the second type of crack mentioned in this Manual classification). Reflection cracking is a separate classification and should not be confused with fatigue cracking.

It can be seen that the details of the defect known as 'cracks' and its various subdivisions in the North American and French standards cover these three subdivisions above, and are clearer, more coherent and more conceptually appropriate.

2.4.4 - Transverse Isolated Crack (NORMA DNIT-005, 2003)

This is an isolated crack that appears in a direction that is predominantly orthogonal to the track centre line. If it is less than 100 cm long, it is called a short transverse crack. If it is longer than 100 cm, it is called a long transverse crack.

2.4.5 - Isolated Longitudinal Crack (NORMA DNIT-005, 2003)

This is an isolated crack that runs predominantly parallel to the centre line of the road. If it is less than 100 cm long, it is called a short longitudinal crack. If it is longer than 100 cm, it is called a longitudinal crack.

2.4.6 - Isolated Shrinkage Crack (NORMA DNIT-005, 2003)

And the isolated crack attributed to thermal shrinkage phenomena, either of the coating material or of the rigid or semi-rigid base material underlying the cracked coating. Comparing the Brazilian standards with the others, this type of crack seems to be associated with Transverse or Longitudinal Cracking outside the wheel tracks.

2.4.7 - Alligator Crack or Fatigue Crack (NORMA DNIT-005, 2003)

This is the set of cracks, with no preferential direction, resembling the appearance of alligator hide. It may or may not be markedly eroded at the edges.

2.4.8 - Interconnected "Block" crack (NORMA DNIT-005, 2003)

This is the set of interconnected cracks characterised by the configuration of blocks formed by well-defined sides. It may or may not show marked erosion at the edges.

2.4.9 - Wear (DNIT, 2006; NORMA DNIT-005, 2003)

Wear is the loss of aggregates or fine mortar from the asphalt surface. It is characterised by abnormal surface roughness, loss of bituminous coating and progressive pulling of the aggregates due to tangential stresses caused by traffic.

This defect can be proved for the following reasons:

J reduction in the bond between the aggregate and the binder due to oxidation of the binder, and the combined action of traffic and weathering agents;

J loss of cohesion between aggregate and binder due to the presence of dust or dirt at the time of construction;

J execution of the work in unfavourable weather conditions;

J the presence of water inside the coating that causes hydrostatic overpressure capable of causing the bituminous film to peel off;

J localised deficiency of asphalt binder in penetration services due to clogged nozzles or poor adjustment of the spreader bar.

As a result of the above probable causes, the asphalt binder is unable to retain the aggregates, which progressively loosen under the action of traffic loads.

It should be noted that this defect is described in great detail in the DNIT Manual (2006), complementing what has already been presented in previous standards.

2.4.10 - Potholes (DNIT, 2006; NORMA DNIT-005, 2003)

Potholes are cavities formed initially in the pavement surface and which vary in size and depth. This defect is very serious because it affects the pavement structurally, allowing surface water access to the inside of the structure. It is also serious from a functional point of view, as it affects longitudinal irregularity and, as a consequence, traffic safety and transport costs.

The main causes of this defect may be related to:

Fatigue cracking (terminal stage);

J localised disintegration on the pavement surface (high severity wear).

It can also occur due to a lack of adhesion between the superimposed layers, causing the layers to peel . Both the onset of this failure and its evolution are accelerated by the action of traffic and climatic factors.

2.4.11 - Patching (DNIT, 2006; NORMA DNIT-005, 2003)

This type of defect can be defined as a portion of the pavement where the original material has been removed and replaced with another (similar or different). It arises from structural or surface problems in the pavement. It occurs for a variety of reasons, usually due to the combined action of traffic and environmental conditions.

It can be deep, when the coating and one or more lower layers of the pavement are replaced, or it can be superficial, when it is only located on the surface of the coating.

2.4.12 - Sinking or Permanent Deformation (DNIT, 2006; NORMA DNIT-005, 2003)

It is characterised by a depression in the surface of the pavement, with or without sill, in the form of:

- J plastic subsidence, caused by the plastic creep of one or more layers of the pavement or subgrade, accompanied by solevation; if its extent is up to 6 m it is called local plastic subsidence, if its extent is more than 6 m it is called wheel track plastic subsidence;
- J Consolidation subsidence caused by differential consolidation of one or more pavement layers or the subgrade, without being accompanied by siltation; if its extent is up to 6 m it is called local consolidation subsidence, if its extent is greater than 6 m it is called wheel track consolidation subsidence.

2.4.13 - Corrugation (DNIT, 2006; NORMA DNIT-005, 2003)

It is a fault characterised by the appearance of transverse, plastic and permanent ripples or corrugations in the asphalt coating.

It can be caused by

- J instability in the bituminous mixture of the coating or base;
- V excess moisture from the underlying layers;
- J contamination of the mixture by foreign materials;
- J water retention in the asphalt mix.

It is most common for this defect to appear in areas where vehicles are accelerating and braking.

2.4.14 - Slipping (DNIT, 2006; NORMA DNIT-005, 2003)

It is a horizontal movement of the surfacing caused by tangential forces transmitted by the vehicle axles (braking and acceleration), which produce a short, abrupt undulation in the pavement in the shape of a half-moon. There is a displacement of the surfacing in relation to the underlying layer of the pavement

It can be caused by

V inadequate connection between coating and underlying layer;

limited inertia of the coating due to its reduced thickness;

V poor compaction;

V plastic creep due to high temperatures.

This defect appears in areas of acceleration and deceleration, such as: uphill and downhill gradients, small radius bends, intersections and bus stops or obstacles.

The North American standards mention the 'slip' defect and the 'corrugation' defect as a single defect to be considered. The French standards only mention the 'corrugation' defect. It can be seen that the classification in these manuals is clearer and more coherent than the DNIT classification. According to the description of these defects in the DNIT Manual (2006), the shape of these defects is slightly differentiated, but in practice this differentiation is barely visible. The causes of these two defects are also detailed in a simpler way in the other manuals (North American and French), and are the same, in the opinion of these technical bodies.

2.4.15 - Exudation (DNIT, 2006; NORMA DNIT-005, 2003)

It is the formation of a film of bituminous material on the surface of the pavement and is characterised by patches of varying sizes. Excess binder appears on the pavement surface, migrating through the coating.

It can occur due to:

S Excess material used in priming or waterproofing;

J inadequate dosage of the asphalt mix, resulting in excessive binder content or very low voids;

J binder temperature above that specified at the time of mixing.

This defect can appear anywhere on the pavement, but it is most common on the wheel tracks.

2.4.16 - Comments

Analysing all the definitions, together with the partial comments made in relation to them, it is possible to conclude that the defect classifications, as presented in the DNIT Manual, can cause difficulties when making decisions, due to the fact that they contradict each other.

Hence the need for greater clarity when defining and detailing each defect, as well as the extent to which the pavement may be affected by it. Only with this clarity can Pavement Management be an effective tool that simplifies the work of road technicians.

In order to make it easier to visualise the defects found in the pavements from the point of view of the three countries, the Comparative Table shown on pages 45 to 52 was drawn up.

In Chapter 3, following on from the comparative table, Defect Assessments are dealt with from the point of view of the three countries, the United States, France and Brazil. Just as the differentiated classification presents some inconsistencies, the assessments are also questioned in terms of their greater or lesser applicability to pavement management in Brazil.

Table 2 - Comparison between the US, French and Brazilian classifications

Defects	**US standards (Obs 1) (appearance / cause)**	**US standards (Obs 1) (severity)**	**US standards (Obs 1) (measurement / repair)**	**French Standards - LCPC, Vizir Method (Obs 5)**	**Brazilian Standards (DNIT Manual, Standard DNIT 005 / 2003 - TER)**
Fissure	Nothing	Nothing	Nothing	Nothing	Aperture visible to the naked eye < 1.50 metres. If > 1.50 m, it is a crack. (Note 3)
Fatigue crack	Irregular alligator-hide blocks, spaced < 0.3 metres apart, appear on the 'wheel track'. Causes: pavement subjected to repeated applications of loads (road traffic) or with loads above the structural capacity	Low (disconnected or lightly connected, no pumping, no edge erosion), Moderate (connections with a defined pattern, perhaps wear and sealing,	The area of occurrence of each severity level is measured. Repair is surface treatment or asphalt slurry (temporary), or patching (permanent). Or	Cracks in the surface (wheel tracks). Develops the appearance of alligator or crocodile leather. It appears due to the continuous action of road	Cracking: cracks, fissures, isolated longitudinal or transverse cracks, interconnected cracks (alligator skin) or block. Classification: FC-1 / FC-2 / FC- 3 (1 - 2 - 3 take into account the increase in severity). Causes: active traffic, by loading and relieving,

	reaches its limit, construction with less than the required thickness or without adequate quality control or improperly conducted rains weaken the pavement structure.	edge erosion, no pumping), High (evident pattern, with sealing and marked wear, with marked edge erosion, pieces removed by traffic, with pumping).	resurfacing. Or pavement reconstruction.	traffic (under repeated loads). There is also a type of circular crack ("faiençage circulaire") which leads to fatigue cracking. **STRUCTURAL DEFECT**	promotes tensile stresses in the inner fibre of the coating; alternation in the daily temperature change which causes contractions in the coating; reflection of existing cracks in cemented bases. There are subdivisions: fatigue (relief and loading), ageing (hardening of the asphalt and loss of flexibility of the binder), reflection. (Obs 2) "
Cracked Blocks	Blocks in the shape of connected rectangles (0.1 to 10 m^2). Causes: contractions in the mix beyond the thermal origin capacity (surface layers with asphalt mix coatings with fine aggregates and asphalt with low penetration), contractions due to variations in moisture content (lower layers), loss of elasticity due to incorrect mixing time or mixtures with excessive temperatures or storage with incorrect time, base with cement or with tropical soils.	Low (up to 6 mm or sealed cracks with undefined dimensions), Moderate (between 6 and 19 mm or cracks less than 19 mm plus a large number of cracks with low severity), High (above 19 mm or a large number of cracks less than 19 mm plus a significant number of cracks with moderate to high severity).	The area of occurrence of each severity level and the standard area of the blocks per section are measured. For points with small dimensions and depths, sealant material is used (asphalt emulsion, surface treatment, asphalt slurry). For more severe areas, patching, recycling, resurfacing or reconstruction is used.	They appear due to shrinkage after the asphalt mixture has been applied.	It is only mentioned under Cracking (indicated under Fatigue cracking in this table) in the DNIT Manual. In DNIT Standard 005, this defect is described as a set of interconnected cracks characterised by the configuration of blocks with well-defined sides, with or without erosion at the edges.
Cracked edges	On unpaved shoulders, at the edge of the shoulder, 60 cm from the edge of the pavement, with irregular and increasing or regular sizes; arises from insufficient compaction or poor drainage.	Low (no cracks or loss of material), Moderate (with some cracks and loss of up to 10 per cent of the pavement), High (cracks with cracks and loss of more than 10 per cent).	The extent of the affected pavement is measured by severity. For repair, sealant is used (to prevent water ingress). If there is poor drainage, it is corrected (perhaps patched).	On the edge of the shoulder with no pavement. Occurs due to inadequate thickness of pavement layers, poor compaction, drainage problems, weather conditions or vegetation.	The DNIT Manual mentions the possibility of longitudinal cracks at the edge of the pavement caused by moisture.
Longitudinal cracks	Parallel to the axle. Inside or outside the wheel tracks. Facilitates water penetration and causes instability	Same as block cracks.	Extent is measured, by severity, both in the wheel tracks and outside the wheel tracks.	It results from imperfect gluing of contiguous strips of pre-mixed. It does	The DNIT Manual describes cracking due to sudden temperature variations that cause thermal shrinkage, especially in the case of

	between pavement layers, fatigue cracks or wear. If it's outside the wheel tracks, it's due to insufficient densification or to the pavement being laid in a different way. If it is at the edge of the wheel tracks, it is due to overloading. It can occur as a result of very hot mixes applied to a fragile base.		The length of cracks sealed with a good sealant is measured. Cracks smaller than 3 mm are not repaired. For cracks between 3 and 20 mm, sealant is used. Cracks over 20 mm are patched, resurfaced or rebuilt.	not always appear in the wheel track. Where the longitudinal crack appears in the wheel track, there may be a 'structural degradation' type defect. **STRUCTURAL DEFECT IF IN THE WHEEL TRACK**	binders with high rigidity. Shrinkage can occur in cemented bases. Depends on temperature variations and the properties of the materials in the mix (does not separate longitudinal and transverse cracking). DNIT 005 only specifies that it is an isolated crack running parallel to the centre line of the road. Up to 100 cm it is called a short longitudinal crack. Above 100 cm it is called a longitudinal crack (Note 8).
Cracks Reflection on Together	Depends on the original crack. It occurs due to discontinuities in the lower layers, which propagate through the surfacing. It arises from: cracks or joints in the lower layers of rigid pavement, movement of bases (treated with cement or lime, or fine sandy lateritic soils), cracks due to low temperature (or block or longitudinal cracks or fatigue cracks) in the old surface layers.	Same as block cracks.	They are measured as longitudinal (length) or transverse (length and amount), by level. Or. as a block, or. by fatigue. The length of the seals is measured. Make: asphalt slurry or. surface treatment, resurfacing or. reconstruction. To prevent spreading, use geomembrane, recycling, others.	Nothing	The DNIT Manual item Cracking mentions that a type of reflection cracking can occur when there is a cracking problem (interconnected or isolated) in the lower layers and this problem propagates to the upper layers.
Transverse cracks	They are perpendicular to the track centreline. They occur due to thermal contraction of the coating and water infiltration into the lower layers.	Same as block cracks	Amount and extension are measured by severity, and extensions of sealed cracks by severity. Lower severity: sealing material. Higher severity: patching, recycling, resurfacing or reconstruction.	It occurs due to: temperature variation (common in countries with cold climates) or degradation of laterite aggregate pavements with a particular shrinkage limit (variation in the water content of the clay fraction).	See comments on the considerations on longitudinal cracks (above), in the case of the DNIT Manual. The DNIT 005 standard only specifies that it is an isolated crack running orthogonally to the centre line of the road. If it is up to 100 cm long, it is called a short transverse crack. If it is longer than 100 cm, it is called a long transverse crack. (Obs 8) "
Patches	It is surface conservation. It is the filling of potholes or any other	Low (< 6 mm barely affecting traffic), Moderate	The occurrences and the area affected by severity are	This is done to regularise the pavement when there are	It is defined as a portion of the pavement where the original material has been removed and

	hole or depression with asphalt mix. A portion of the surface pavement, longer than 0.1m^2, which is removed and replaced with material applied over the original construction of the road. Various defects cause the need for patching.	(between 6 and 12 mm moderately affecting traffic), High (> 12 mm and with occasional pumping and serious interference with traffic). It is measured by severity.	recorded. The area is cut out and cleaned, the waterproofing and adhesive layer is applied and the replacement material is applied. If the material is applied without cutting out the area, it is called a pothole cover.	surface or deep defects. If they are frequent, there may still be a problem. **STRUCTURAL DEFECT - IF NOT COMPLETELY CORRECTED**	replaced with another (similar or different). It arises from structural or surface problems with the pavement. They can occur for a variety of reasons, usually due to the combined effects of traffic and environmental conditions. It can be deep (the surfacing and one or more lower layers are replaced) or superficial (the surface of the surfacing is corrected).
Pots or Holes	Cavity (like a 'bowl') that appears in the surfacing, which may or may not reach the underlying layers. It results from local disintegration under the action of traffic and water; with fragmentation and removal of the pavement after cracking due to fatigue or wear. It appears in coatings with little thickness or low bearing capacity (structural failure). It can appear in places with material segregation (lack of binder) or drainage problems.	Low (up to 25 mm deep), Moderate (25 to 50 mm deep), High (over 50 mm deep).	The number of occurrences and the relative area affected by level of severity should be recorded. Repairs should be made by patching, followed by resurfacing if necessary, depending on the size of the affected area.	It appears as a result of wear or destruction of the bearing layer. Initially, its dimensions are small. If not corrected, its dimensions can increase, leading to the removal of the pavement. It starts in the surfacing, but can reach the base if not repaired.	Cavities formed in the pavement covering and varying in size and depth. Structurally affect the pavement, allowing surface water access to the interior of the structure. They affect longitudinal irregularity and, as a consequence, traffic safety and transport costs. causes can be: fatigue cracking (terminal stage); localised disintegration on the pavement surface (high severity wear). The process can be accelerated by traffic and the weather.
Permanent Deformation	It appears in the path of the 'wheel tracks'. It is a longitudinal subsidence that follows the path travelled by vehicle wheels. It occurs due to densification of the materials or shear rupture. Causes: inadequate sizing of layer thicknesses, inadequate dosage of the asphalt mix, inadequate compaction and subsequent consolidation by traffic, drainage failure. If there is shear failure, lateral elevations may appear parallel to the traffic.	Low (up to 25 mm deep), Moderate (25 to 50 mm deep), High (over 50 mm deep). **SEPARATION BY FRENCH STANDARD (similar and cited for comparison) Gravity varies between weak (up to 2 cm in height), medium (2 to 4 cm in height), strong (above 4 cm). 4 cm).	Measurements of the gradient or maximum depths of deformation, in millimetres, should be noted at every 20m (metre) interval. The following can be used to repair this defect: recycling, resurfacing or total reconstruction of the pavement.	It appears 50 - 80 cm from the edge due to: insufficient support (or low capacity) or poor quality, poor drainage (varies with the base), settlement of materials under heavy and channelled traffic associated or not with a drop in resistance of the lower layers. There is 'localised or punctual subsidence' ("affaissement hors rive" or	It is a depression in the pavement surface, with or without sill, taking the form of: - plastic subsidence (with sill) caused by the plastic creep of one or more layers of the pavement or subgrade (up to 6 m is local plastic or over 6 m and in the wheel track is wheel track plastic); - consolidation subsidence (no solevelling) caused by differential consolidation of one or more pavement layers or the subgrade (up to 6 m is local consolidation or over 6 m and in the wheel track is wheel track consolidation). (Obs 8)

				"flache"), with the same origin but small area. **STRUCTURAL DEFECT**	
Ripple or Corrugation or Slippage	Longitudinal displacement of the pavement, causing transverse undulations. It is caused by 'braking' or 'acceleration' movements. It arises due to: inadequate dosage of the asphalt mix, poor bonding between base and surfacing, insufficient structure. It is located at curves, intersections, slopes or gradients. (Obs 4)	Severity levels can be determined according to the extent to which traffic quality is affected.	Measurement is carried out by recording the number of occurrences and the area affected (in m^2). Patching, recycling, resurfacing or reconstruction must be carried out.	Occurs when the pavement is under heavy stress (+ on bends or ramps or in places subject to braking and acceleration). This classification can include occasional defects ("gonfle"), not always in places with acceleration and braking, but with similar characteristics ("stewing" in the pavement). It is caused by chemical incompatibilities or poor compaction.	Corrugation: transverse (permanent) undulations in the surfacing. Causes: instability in the asphalt mixture of the surfacing or base; humidity in the underlying layers; contamination of the mixture by foreign materials; water retention in the mixture. Occurs with acceleration and braking. Slipping: horizontal movement of the surfacing, due to tangential forces transmitted by vehicle axles; produces short, abrupt undulations in the pavement (like a half-moon). Causes: inadequate bond between surfacing and underlying layer; limited inertia of the surfacing due to reduced thickness; poor compaction; plastic creep due to
					high temperatures. Occurs when accelerating and braking (uphill and downhill, small radius bends, intersections and bus stops or obstacles). (Note 7)
Exudation	Excess bituminous material on the surface, usually in the wheel track. It appears as asphalt with loss of normal colour or loss of common texture. The aggregate shows excess bituminous material (darker colour). The surface is shiny and sticky. Tyre marks appear with high temperatures and reduced tyre-pavement friction. The result of inadequate asphalt mixing (low voids).	It is classified according to the situation or the evolution of the problem. Low: there is a change in the colour of sections of the pavement compared to the general situation. Moderate: there is a change in the texture of the pavement. High: the pavement has a shiny appearance and tyre marks appear on it (in hot weather).	To quantify this type of defect, the area of occurrence is checked and classified by level of severity. It should be repaired with surface treatment or hot sand, or even recycled.	Excess binder appears on the surface.	It is the formation of a film of bituminous material on the surface of the pavement and is characterised by patches of varying sizes. It occurs due to: inadequate dosage of the asphalt mix, which results in excessive binder content or a very low voids index, binder temperature above that specified at the time of mixing. It can appear anywhere on the pavement, but preferably in the wheel tracks.

Polished Aggregates (Obs 4)	Polishing or wearing of the aggregates and bituminous binder can expose the coarse aggregate. Loss of material can occur. This defect can be the result of the abrasive action of traffic, which removes asperities and angularities from the particles.	There is no way of defining severity levels. The degree of polishing can reduce tyre-pavement friction and compromise road safety.	The affected areas are measured in metres^2. Surface treatment or asphalt slurry should be carried out. Or recycling, resurfacing, pavement reconstruction.	Nothing on the record.	Nothing on the record.
Wear and tear	It arises from the displacement of aggregate particles and loss of adhesion of the binder. It evolves from loss of fines to loss of some coarse aggregate, and finally to major losses of aggregate. Causes: ageing, hardening, oxidation, volatilisation, weathering. To ensure good adhesion between aggregate and	They can be classified according to the local situation or the evolution of the problem. Low level is when there is a loss of fine aggregates. Moderate level is when the surface texture becomes rougher and some coarse aggregates are lost. High level is	Measurement is done by quantifying the area of occurrence, classified by level of severity. Repairs can include: sealcoating, surface treatment or asphalt slurry, recycling or resurfacing.	Initially, there is little loss of material. With the passage of time, this loss accelerates and jeopardises the quality of the pavement.	Loss of aggregates or fine mortar from the coating. It shows abnormal surface roughness, loss of bituminous coating and gradual pulling out of aggregates. Causes: reduced adhesion between aggregate and binder (due to oxidation of the binder + combined action of traffic and weathering agents, presence of dust or dirt in the construction), work carried out under poor conditions.
	binder, avoiding the likelihood of early wear, the following must be carried out: proper asphalt mixing (dosage or temperature of execution); application in good weather conditions, correct compaction and adequate time for release to traffic. (Obs 4) "	when the surface texture becomes very rough and there is a more intense loss of coarse aggregates.			poor weather conditions, the presence of water in the coating (hydrostatic overpressure can cause the bituminous film to peel off), asphalt binder deficiency due to clogged nozzles or poor adjustment of the spreader bar. For these reasons, the asphalt binder is unable to retain the aggregates that loosen under the action of traffic loads. (Well detailed in the DNIT Manual and complements what was presented)
Gap between carriageway and hard shoulder	Difference in elevation between the carriageway and the shoulder. It results from the application of different layers of material (sometimes a greater thickness is applied to the road surface than to the shoulder). In the case of an unpaved	Severity levels are not applied in this case.	Gap measurements are taken every 20 m (metres). In order to repair this defect, the shoulder must be replaced at the site of the incident.	Paved shoulder: unevenness between the shoulder and the carriageway, causing a dangerous step. With an unpaved shoulder: erosion on the shoulder	Nothing on the record.

	shoulder, there may be erosion on the shoulder, causing a gap between the shoulder and the carriageway. There may be differential consolidation between the carriageway and the shoulder.			jeopardises traffic safety (it can reach the carriageway). The severity of this defect can only be determined here: low - gradient of 1 to 5 cm, medium - gradient of 5 to 10 cm, high - gradient > 10 cm. (Note 5)	
Pumping (Obs 4)	Leakage or ejection of water from the underside of the pavement through cracks when under the action of traffic. It is detected by the presence of 'fines' on the surface (eroded from the support or base layers). Occurs if there is water under the surface (drainage deficiency).	It's difficult to define severity levels because the amount of water pumped can vary according to humidity conditions.	The number of incidents and the area affected are recorded. If the pavement has been damaged, it must be repaired. The drainage system at the site must be corrected to prevent further pumping.	Nothing on the record.	Nothing on the record.
Peeling (Obs 6)	Nothing reported (include in the hole or pothole item, even if with low severity).	Nothing listed (include in hole or pot).	Nothing listed (include in hole or pot).	Parts of the bearing layer come loose (detach) from the base.	Nothing listed (include in hole or pot).
Combing (Obs 6)	No record (include in wear item).	No record (include in wear item).	No record (include in wear item).	Poor execution of the asphalt layer due to uneven distribution of the binder (nozzle failure), or incorrect distribution of aggregates.	No record (include in wear item).

Note 1: The ways of identifying defects, their possible causes, classification by severity levels (including how to quantify this severity) according to the FHWA (1993), complemented by the works of Fernandes Jr. et al. (2003) and Haas et al. (1994), have been cited. The levels of severity and how to measure each defect are described only in relation to these works, since they are described more precisely in these bibliographies (terms such as 'clearly open cracks' used in the French standards can give rise to different interpretations). Pavement repair methods are similar in all three countries.
Note 2: The DNIT Manual does not detail how each type of crack or fissure or crack appears and what can cause each occurrence. This item in the table combines the DNIT Manual (Cracking) and Standard DNIT 005 / 2003 - TER (item Alligator Crack). Further details of the various types of cracks can be found in Standard DNIT 005 / 2003 - TER and are referred to as block cracks, longitudinal cracks and transverse cracks. Reflection cracks and edge cracks are mentioned superficially in the DNIT Manual. Shrinkage cracks

(mentioned in DNIT 005 / 2003) are not mentioned, considering that they can be classified as longitudinal cracks or transverse cracks. The division into fatigue cracks, ageing cracks and reflection cracks, mentioned at the end of this item, is confusing and even changes concepts that are already recognised as appropriate, as mentioned in item 2.4.3 (page 38).
Note 3: Brazilian standards subdivide cracks (any discontinuity in the pavement surface) into fissures and cracks. 'Cracks' was cited as the first item in the table. As cracks are defects only mentioned in Brazilian standards and do not require programming for repair, they are not considered for pavement management purposes.
Note 4: It is possible to observe a classification by US agencies that is similar to corrugation: it is "slippage cracking" or "parabolic cracking", which occurs in acceleration areas of vehicles: when accelerating, the vehicle goes forwards and the friction force 'pushes' the pavement backwards (it slips). Within the concept of 'wear', there is a subdivision in some US agencies that classifies two types of this defect. One is "stripping", which is a type of wear that occurs if there is incompatibility of materials and, under the action of water, wear begins (early), the other is "wear loss", which is a dislodgement of aggregates due to ageing and loss of adhesiveness (occurs with the age of the pavement). Polished aggregates and pumping are only mentioned in US standards, but it is important to evaluate them in view of their constant occurrence in Brazilian pavements.
Note 5: The French standards do not detail the causes, ways of measuring and severity levels of some defects. Only these standards mention levels of severity of the Gap between carriageway and shoulder, and these levels should be considered for a future general classification.
Note 6: Peeling is indicated as a defect only in the French standards, and can be included in the Hole or Pan defect. Combing is indicated as a defect only in the French standards, and can be included in the Wear defect.
Note 7: The North American standards mention the 'slip' defect and the 'corrugation' defect as a single defect to be considered. The French standards only mention the 'corrugation' defect. It can be seen that the classification in these manuals is clearer and more coherent than the DNIT classification. According to DNIT's description of these defects, the shape of these defects is not very distinctive and is not very visible in practice. The causes of these two defects are also detailed in a simpler way in the other manuals (North American and French), and are the same, in the opinion of these technical bodies. The places where these defects appear are also the same.
Note 8: It is better and more important to quantify the length of the crack than to set limits for the type of crack, such as the limit value of 100 cm for a longitudinal or transverse crack to be called short or long. The same applies in the case of permanent deformation, which uses the limit value of 6 m for the deformation to be called local or wheel track.

CHAPTER 3

DEFECT ASSESSMENTS ON FLEXIBLE PAVEMENTS

3.1 Initial considerations

It is important that both comfort and safety aspects are taken into account so that a road can be assessed as being of good quality or not. A comfortable and safe road will consequently be a road with more economical traffic.

Evaluating the defects that appear in pavements allows them to be repaired so that users can have their safety and comfort restored as quickly as possible. In order for the necessary repairs to be carried out in the most appropriate way and at the right time, with the best use of available financial resources, this assessment must be as efficient as possible.

With this in mind, both subjective and objective carriageway assessment methods have been developed. Subjective evaluation is based on users' opinions of the road in terms of the comfort with which drivers travel on it, identifying how satisfied or dissatisfied drivers are with the road surface by assigning scores or concepts. The objective assessment, on the other hand, indicates the extent to which the road's rolling quality and safety are jeopardised by pavement irregularities, and for this purpose these irregularities are measured using various methods.

3.2 Flexible Paving Reviews in the United States

The initial reference for assessing defects in flexible pavements in the United States was the AASHO (American Association of State Highway Officials) Road Test (CAREY and IRICK, 1960), where subjective and objective assessments were made.

3.2.1 - Subjective evaluation

A good level of service on a road is based on characteristics that provide pleasant, comfortable and safe traffic. Information about the current situation of the road must indicate what this level of service is at the time of its assessment, so that decisions can be made about interventions on the road in an appropriate manner.

On the occasion of the AASHO Road Test (1956-1961), the concept of 'serviceability' was developed by Carey and Irick (1960). This concept defines that 'the level of service is a measure of the road's ability to serve traffic at a given time'. The level of service decreases with time or the useful age of the road (or both) and also as a result of the natural wear and tear that the pavement suffers from the action of road traffic. This variation is defined as pavement performance.

The subjective method used by Carey and Irick (1960) consists of a group of evaluators assigning a score, resulting in an average value of the level of usefulness or Current Use Value (USV). The scores range from 0 (very bad) to 5 (very good). The section considered must also be assessed as 'acceptable' (yes) or 'not acceptable' (no), or the option 'doubtful' (undecided).

Thus, considering the variation of this level of serviceability, or performance of the road pavement, over the useful life of the road, it is possible to draw a curve of variation of the PSI (Present Serviceability Index) or VSA (Present Serviceability Value) as a function of Time or Traffic.

Figure 22, based on Femandes Jr. et al. (2003), illustrates the concept of pavement performance.

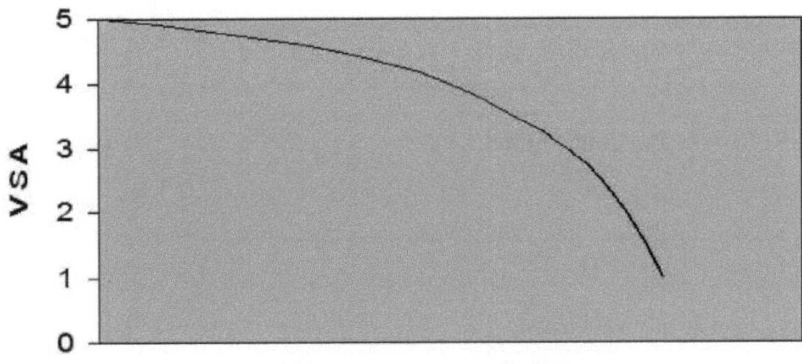

Figura 22 - Performance Concept: Variation of VSA with Time or Traffic Source: Femandes Jr. et al. (2003)

The level of serviceability attributed to the current condition of the road results from the average of the evaluations, with 4 to 5 being the range attributed to a new pavement, while the range of 1.5 to 2.5 corresponds to a pavement on the edge of needing repairs. Figure 23, taken from Femandes Jr. et al. (2003), illustrates the minimum acceptable threshold for making pavement repairs.

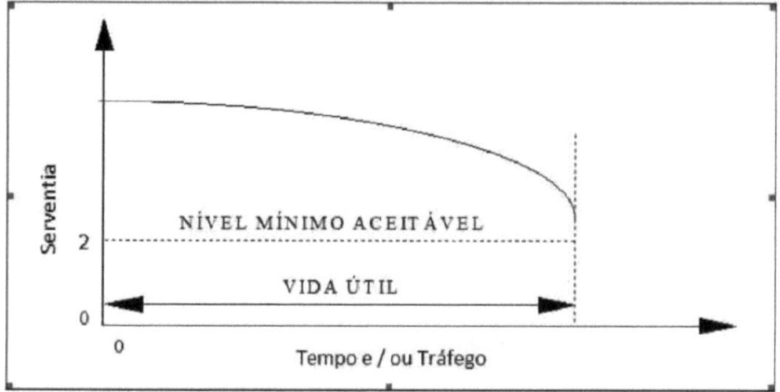

Figura 23 - Need for intervention based on the concept of service-performance

Source: Femandes Jr. et al. (2003)

Several attributes are taken into account to 'qualify' the pavement, in particular:

J how satisfactorily vehicle and track are interacting at the speed considered;

-How bad the appearance of the road is in terms of cracks, potholes, texture and colour.

3.2.2 - Objective Assessment

While the subjective evaluation expresses the opinion of the evaluators regarding the road pavement, the objective evaluation measures the defects that affect the pavement's serviceability.

The longitudinal irregularity of the pavement is the property most easily perceptible to users. An example of irregularity can be seen in Figure 24, taken from the DNIT 005/2003-TER Standard.

Figure 24 - Longitudinal Irregularity Source: DNIT Standard 005/2003 - TER (2003)

The longitudinal irregularity of pavements is defined as distortions in the pavement surface in relation to a reference plane.

The road's bearing capacity can also be assessed, i.e. whether the pavement is capable of withstanding the loads it will be subjected to during its useful life. Another form of assessment related to road safety is the determination of the coefficient of friction between the road and the vehicle, and whether or not it is adequate.

There is equipment that assesses longitudinal irregularity, bearing capacity and friction between the road and the vehicle.

Profilometers, profilographs and response-type equipment can be used to assess *longitudinal irregularity*.

Profilometers measure the angular variations between the plane in question and the road surface (Figures 25 and 26). *Profilographs* measure the variations in road surface elevation between the various wheel assemblies that support the equipment's structure (Figure 27). Figure 25 was taken from Femandes Jr. et al. (2003).

Figure 25 - CHLOE profilometer, used in the AASHO Road Test

Source: Femandes Jr. et al. (2003)

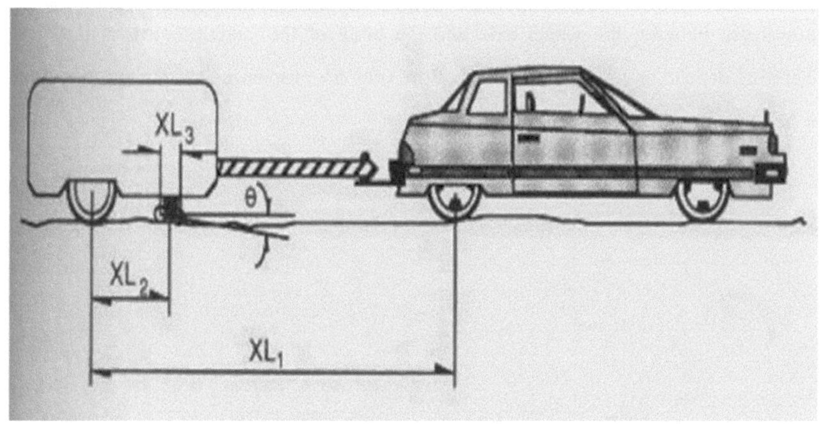

Figura 26 - Schematic of the CHLOE profilometer

Source: Haas et al. (1994)

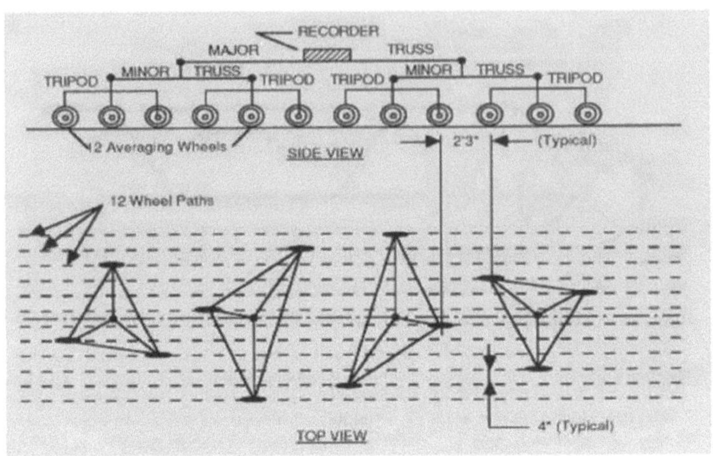

Figura 27 - Schematic of a Profilograph

Source: Haas et al. (1994)

Response-type evaluation equipment quantifies longitudinal irregularity indirectly by measuring the differential movement between the wheel axle and the body of the vehicle being used. These evaluations therefore depend on the characteristics of the vehicle's suspension system and travelling speed. Figure 28 illustrates this equipment.

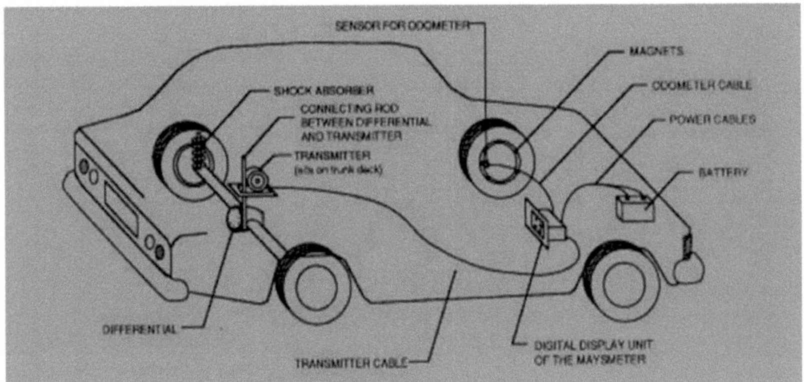

Figura 28 - Maysmeter: response-type irregularity meter

Source: Haas et al. (1994)

Accelerometers can also be used to quantify the vertical acceleration ('jolts') to which the vehicle is subjected while travelling over the pavement being analysed. Figure 29 illustrates a vehicle equipped with an accelerometer and is taken from Femandes Jr. et al (2003).

Figura 29 - Car Road Meter: equipped with an accelerometer developed by the FHWA
Source: Femandes Jr. et al. (2003)

Structural, destructive or non-destructive tests can be carried out to assess the *bearing capacity of* a pavement.

In the first case, *samples of the pavement* are taken for laboratory testing. In the second case, pavement deflection measurements are made using the following devices: *Benkelman beam, vibrating deflectometers, impact deflectometers* (FWD - Falling Weight Deflectometer). Both the Benkelman beam and the FWD can provide the deflection basin of the evaluated section. The deflection basin consists of plotting the values of the elastic or recoverable displacement measurements at various points from the centre of loading (point of load application). As such, just below this centre are the largest values, and as you move away from the centre, these displacements become smaller and smaller, until they are negligible. Figure 30, taken from Bemucci et al. (2007), illustrates the deflection basin.

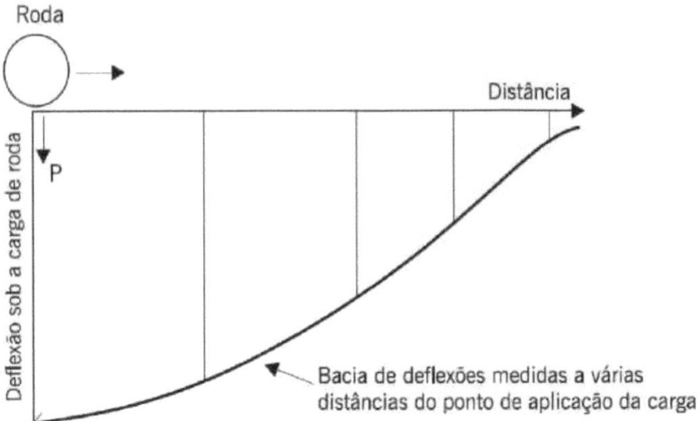

Figura 30 - Diagram of readings to obtain the deflection basin

Source: Bemucci et al. (2007)

The Benkelman Beam is a piece of equipment that is used quite frequently, and was even used for the structural assessments carried out by the DER on the section of the Case Study. This device has a fixed part and a mobile part. The fixed part is supported on the pavement by three adjustable feet. The mobile part is coupled to the fixed part via a joint, one end of which is in contact with the pavement (test tip) and the other end of which is in contact with the strain gauge that detects any vertical movement of the test tip. The fixed part is fitted with a vibrator to reduce friction between all the moving parts during operation.

The standard that prescribes the test method for determining pavement deflections using the Benkelman Beam is DNER-ME 24/94. Figures 31 and 32, taken from the DNER-ME 024/94 standard (1994), show, respectively, the Benkelman beam schematic and the reference schematic on the beam and the lorry carrying it.

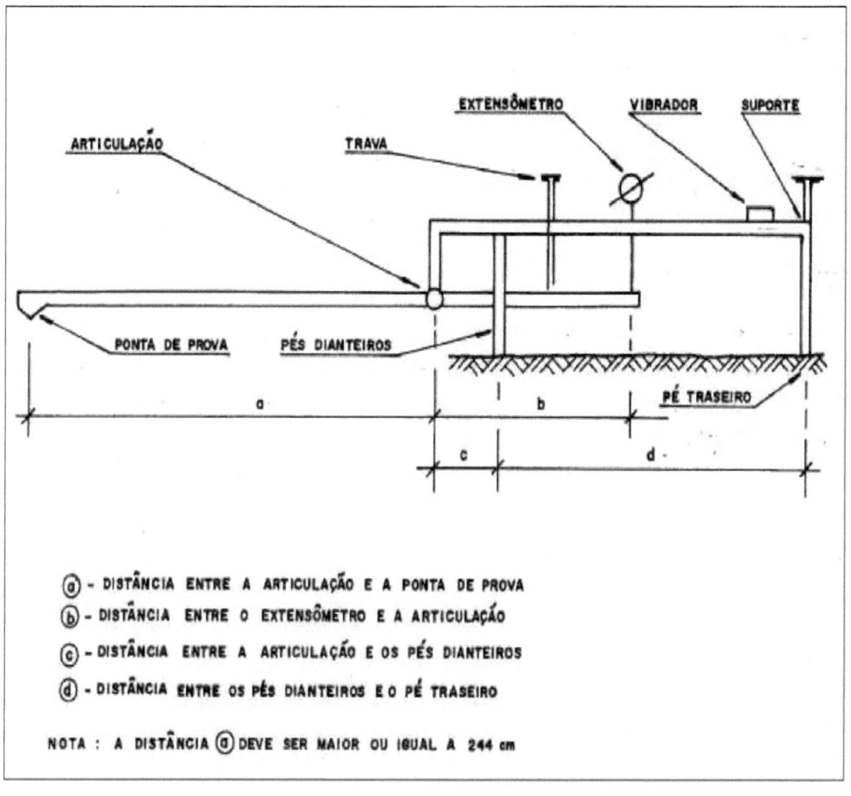

Figura 31 - Benkelman Beam diagram

Source: Standard DNER-ME 024/94 (1994)

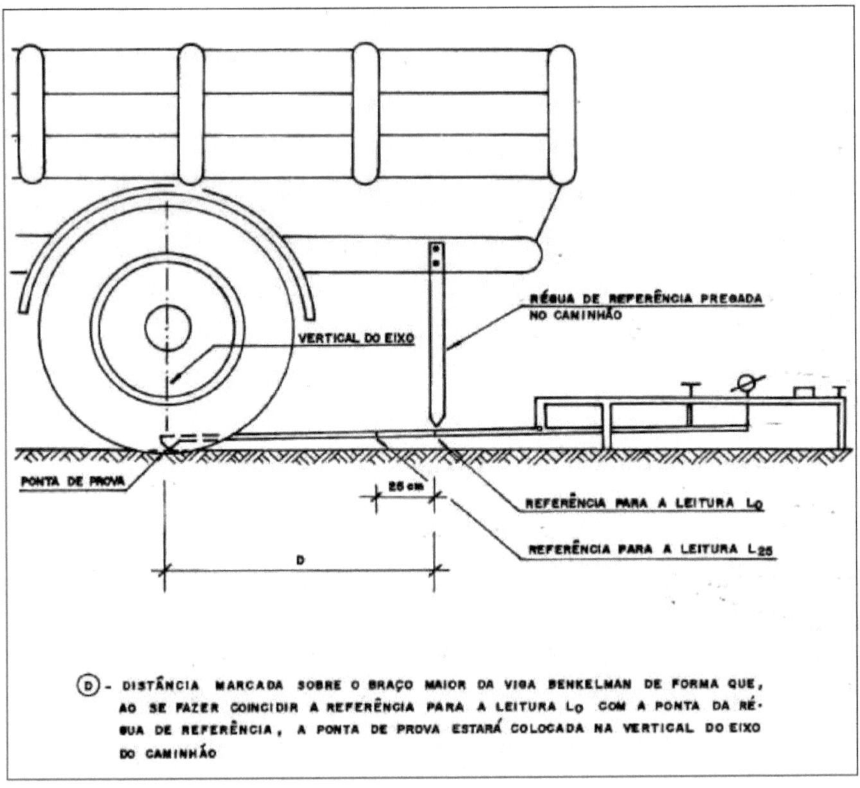

Figura 32 - Reference diagram on beam and lorry

Source: Standard DNER-ME 024/94 (1994)

To assess *road safety conditions,* you can check the *tyre-pavement surface friction*, which can also tell you whether a particular road segment has enough friction for traffic to be safe on that stretch. There is *equipment, such as the Mu-Meter*, which measures the friction forces on a trailer travelling with the wheels locked, at different speeds, on a wet pavement. Figure 33, taken from Bemucci et al. (2007), shows a detail of this type of equipment.

Figura 33 - Example of Mu-Meter friction measuring equipment

Source: Bemucci et al. (2007)

3.2.3 - Aspects of US evaluations relevant to Brazil

Appraisals carried out in the USA are aimed at detail and precision, always using the most modern equipment. This type of procedure can often be expensive and delay the decision-making process. However, the reliability of the SHRP has proven to be adequate in cases where greater precision is required. Therefore, in many cases, when assessing defects in Brazil, some guidelines from the US standards could be used when a greater margin of safety is required.

3.3 Flexible Pavement Assessments in France

Between 1965 and 1967, road assessments were carried out in France in order to define an action plan to remedy the problems caused by the harsh winter of 1962/1963, which had a major impact on French roads. Later, between 1970 and 1978, coordinated reinforcement studies were carried out to restore the road network. During this period, the catalogue of pavement degradation (1972) and the flexible pavement survey guide (1977) were published.

As a result of the new publications, there was concern about the implementation of support systems for road maintenance management, within which the VIZIR assessment method met this new objective.

The approach taken by the French standards is different, especially because two types of degradation are defined: type A, which is linked to the bearing or structural capacity of the road surface, and type B, which is linked to causes other than this (defects in construction or in the quality of a product, particular local conditions that traffic can accentuate, etc.).

As the 'French manuals' adopt this type of classification, it is necessary to be clear about the origin of the defect being treated. By knowing whether or not there is a structural problem, it is possible to better define the most appropriate type of intervention for each case.

In the case of type B degradation, the choice of type of intervention (solution to the problem) derives from recognising the degradation alone, and there is no need to use other parameters to make the diagnosis. In the case of type A degradation, the solution will depend on many factors and the diagnosis will combine degradation, support and traffic. Thus, type B degradations are only analysed when type A degradations are absent.

In addition, two parameters need to be defined:

-*S* extent (length, area, number of occurrences); severity (low, medium, high).

3.3.1 - Surface degradation index (SI)

The degradation index, also known as the global visual index, is calculated over a length of road, based on three groups of degradation: cracking (which leads to the structural defect 'fatigue cracks'), deformations (or wheel tracks) and patches.

Figure 34, taken from Bemucci et al. (2007), illustrates a fatigue crack with erosion (hole or pot initiation).

Figure 34 - Fatigue cracks with erosion

Source: Bemucci et al. (2007)

According to VIZIR, the IF or 'cracking index' is calculated, which is defined by the severity and extent of cracking or fatigue cracking. Then the ID, or 'deformation index', is calculated, which is defined by the severity and extent of permanent deformation. And finally, the two indices are combined and corrected by the 'repair rate', which is determined by the severity and extent of patching.

Through this procedure, the IS is obtained, which is the 'surface degradation index', or also the 'visual state'. This index ranges from 1 to 7. Scores 1 and 2 correspond to good condition. Grades 3 and 4 correspond to average condition. Grades 5, 6 and 7 correspond to very poor condition, requiring major conservation or reinforcement work.

It should be noted that this classification of indices between 1 and 7 is unnecessary, given that the visual states are then grouped into just three: good (1 and 2), average (3 and 4) and bad (5 to 7).

3.3.2 - Support capacity

By measuring the 'deflections' in the pavement, the 'bearing capacity' of the pavement is assessed. Deflections are measured on the sections analysed, and three categories are classified: up to **dl** the pavement behaves well, beyond **d2** the pavement behaves badly, between these limits is the intermediate region.

To set these limits, various factors are taken into account, such as the climatological environment, the nature and thickness of the pavements, the soils and the axle loads. It is also advisable to take into account the volume of traffic on site when determining the **dl** and **d2** limits, with the greater the traffic the lower the permissible deflection.

3.3.3 - Aspects of Valuation in France that are relevant to Brazil

The Assessment of the *Surface Degradation Index (Visual State)* and the *Carrying Capacity* provide input for 'decision-making', mainly 'at the network level', i.e. 'at the macro level'. These analyses are intended to cover objective and subjective aspects in a complementary way.

For final decision-making and definitions of 'specific solutions for the various sections analysed' ('project-level' decision-making), the VDM (average daily traffic volume) of the road must also be taken into account. For higher VDMs, the tolerance for intervention is lower and evaluation should take place at shorter intervals.

According to Prestes (2001), the time taken to apply the methods using SHRP was longer than VIZIR, which can be explained both by the greater number of defects assessed and by the fact that measuring defect areas takes longer. However, the reliability of the SHRP has been shown to be greater, with more detailed and accurate results.

3.4 Flexible Pavement Evaluations in Brazil

The pavement assessments used in Brazil and detailed in specific standards are based on US standards, with only a few specific guidelines for the Brazilian case.

DNIT 005/2003-TER, a revision of DNER-TER 001/78, defines the defects to be assessed. The other procedural standards were drawn up on the basis of DNIT 005/2003-TER. DNIT 009/2003-PRO, a revision of DNER-PRO 007/94, is used for subjective assessment. DNIT 006/2003-PRO, revision of DNER-PRO 008/94, and DNIT 007/2003- PRQ, revision of DNER-ES 128/83, are used for the Objective Assessment. DNIT 008/2003-PRO is used to assess the pavement surface using the

Continuous Visual Survey.

3.4.1 - Subjective evaluation

DNIT Standard 009/2003-PRO standardises certain procedures so that a subjective assessment can be carried out on an appropriate basis. It is used at federal, state and municipal level, where appropriate.

The VSA or Current Service Value is determined through subjective evaluations carried out by a group of evaluators who travel the stretch under analysis, recording their opinions on the pavement's ability to meet the demands of the traffic on it at the time of the evaluation, in terms of smoothness and comfort. This survey must comply with certain guidelines:

As for the people (evaluators), they must be sensitive to the work to be done;

the sections should be of a fixed length, uniform in appearance and preferably small enough to be evaluated in a short space of time;

a standard form should be used;

The stretch should be assessed as if it were a motorway with heavy traffic and for commercial or passenger vehicles;

the aspect to be considered is only the current one, even if there is an apparent possibility of rupture in the future;

-The evaluation should be carried out under favourable weather conditions;

the geometric aspects of the road, railway crossings, bridge accesses, culvert embankments, etc. should not be taken into account;

mainly potholes, bumps, transverse and longitudinal surface irregularities should be considered.

The VSA is the sum of the scores awarded by the evaluators divided by the number of evaluators. This assessed value is always for the specific moment analysed.

The grades are related to the concepts below:

Bad (between 0 and 1)

J Bad (between 1 and 2)

-*S* Regular (between 2 and 3)

J Good (between 3 and 4)

J Good (between 4 and 5)

These grades will be used to make the necessary decisions, and it is advisable to consider the 'objective assessment' together.

It should be emphasised that Carey and Irick (1960) avoided using the terms 'very bad', 'poor', 'excellent', etc. They preferred to use simple and direct terms such as 'very good' or 'very bad'. This option allows us to think in 'ranges' because the grades cover more than one value (0-1 and 4-5) and not absolute values. The Brazilian standard's choice to use the terms 'very bad' and 'excellent' could give rise to different interpretations.

3.4.2 - Objective Assessment

Standard DNIT 006/2003-PRQ, later supplemented by Standard DNIT 007/2003- PRO, standardises certain procedures for determining the apparent occurrences and permanent deformations in wheel tracks.

The occurrences to be recorded are defined in the tables of the aforementioned standards. In general terms, the following are assessed: cracks, subsidence (plastic or consolidation), rippling or corrugation, slipping, exudation, wear, pans or holes. Cracks can be isolated (transverse or longitudinal) or interconnected (blocky or with the appearance of alligator skin). These defects were detailed in Chapter 2.

Stations to be surveyed (or analysis sites) are marked out. The length to be analysed at each station is 3.00 m before and 3.00 m after. These stations are marked every 20 metres alternating with the axis of the carriageway (every 40 metres in each traffic lane). The surfaces to be assessed must be demarcated using demarcation paint (the thicknesses of the strokes are detailed in the standard) and clearly numbered to avoid doubts at the time of the survey.

The aim is to determine the deflection in the wheel track using a standardised aluminium lattice. At the same time, all other defects are identified, delimited and recorded.

All standard recommendations must be strictly adhered to in order to guarantee a standardised procedure. Two parameters are identified at the end of the work:

fr, which is the relative frequency (obtained using the formula **fa** x **100 / n,** where **fa** is the number of times the occurrence was verified and **n** is the number of inventoried stations or analysis sites)

J **fp,** which is the weighting factor (tabulated in the standard, varying according to the defect assessed)

It is important to note that the weighting factors for the most serious interconnected cracks, sinkholes, potholes, patches, corrugations and slippage have higher values. With the exception of block cracks (also considered in the 'interconnected cracks' item), corrugations and slippage, the other items are the defects considered, in the French standards, to be defects of structural origin. In the case of these standards, they are the values with the greatest weight when evaluating the pavement. The values of the weighting factors, taken from Standard DNIT 006/2003-PRO, are in Annex 2, on page 108.

The **fr** x **fp** ratio gives the **IGI** (individual severity index) for each defect and the **IGG** (global severity index), which is the sum of the partial **IGIs**. These values are associated with the concepts below:

Bad (IGG > 160)

J Bad (80 < IGG < 160)

J Regular (40 < IGG < 80)

J Good (20 < IGG < 40)

J Optimal (0 < IGG < 20)

With the above figures in mind, decisions can be better informed. It's worth remembering, once again, the issue of the terms 'bad' and 'excellent', already mentioned in section 3.4.1..

3.4.3 - Continuous Visual Survey

Standard DNIT 008/2003-PRO establishes the conditions for evaluating the pavement using the Continuous Visual Survey process, determining the ICPF (Flexible Pavement Condition Index) and providing elements for calculating the IGGE (Expedited Global Severity Index). Based on these indices, the IES (Pavement Surface Condition Index) is calculated for each section under consideration.

To calculate the ICPF, all defects detected during the LVC process are taken into account. To calculate the IGGE, the defects 'cracks', 'permanent deformations', 'pans' and 'patches' are taken into account. For the purposes of the LVC, the definitions and nomenclatures of Standard DNIT 005/2003-TER are used, which can be found in Annex 1 (page 107) of this work.

For the continuous visual survey, a vehicle equipped with a calibrated speedometer/odometer should be used to measure operating speed and distances travelled. Rainy days, foggy days or days with little natural light (early or late in the day) should be avoided. In addition to the driver, there must be at least two technicians for this survey. The average working speed is 40km/h. The stretches to be assessed must be divided in such a way as to preserve some homogeneity. To carry out the survey, a form set out in the aforementioned standard is filled in.

All the detailed instructions for calculating the ICPF, recommended in DNIT Standard 008/2003-PRO, must be strictly adhered to in order to guarantee a homogeneous standard of procedure.

In the case of the IGGE, only a few defects are taken into account, and the formula for the calculation is:

IGGE = (Pt × Ft) + (Poap × Foap) + (Ppr × Fpr), onde:

where:

- *A* Pt and Ft = weight and frequency of the crack set;
- Poap and Foap = weight and frequency of the set of deformations;
- *A* Ppr and Fpr = weight and frequency of the set of pans and patches.

Defects such as cracks, deformations and pans or patches are assigned weights according to their severity. The lower weights are for cracks and the higher weights are for pans and patches. It is important to note that some of the defects used to calculate the IGGE are those that the French standards indicate as being defects of structural origin (cracks, which include fatigue cracks, deformations and patches). After calculating the ICPF and IGGE, the IES will be determined, according to Table 3 (DNIT 008/2003-PRO):

Table 3 - Variation in IGGE values.

Description	HEI	Code	Concept
IGGE<20 and ICPF>3.5	0	A	Great
IGGE<20 and ICPF<3.5	1	B	Good
20<IGGE<40 and ICPF>3.5	2	B	Good
20<IGGE<40 and ICPF<3.5	3	C	Regular
40<IGGE<60 and ICPF>2.5	4	C	Regular
40<IGGE<60 and ICPF<2.5	5	D	Bad
60<IGGE<90 and ICPF>2.5	7	D	Bad
60<IGGE<90 and ICPF<2.5	8	E	Bad
IGGE>90	10	E	Bad

3.4.4 - Considerations on valuations carried out in Brazil

As you can see, assessments based on DNIT standards cover three possibilities: subjective assessment (3.4.1), objective assessment (3.4.2), assessment with both approaches (3.4.3).

The latter is the LVC (DNIT Standard 008/2003-PRO), which combines the IGGE (objective analysis) with the ICPF (a slightly more subjective analysis).

In some places, in addition to objective and subjective evaluations, it is advisable to carry out a survey of drivers' general understanding of pavement conditions. A study like this was carried out by Danieleski (2004) on the road system in Porto Alegre, where the usefulness of applying joint assessments was verified.

3.5 Objective Evaluation with Determination of the Pavement Condition Index (PCI)

Once the defects have been identified, it is necessary to define which combinations of the extent and severity of the different defects indicate that one pavement section is in worse condition than another. To this end, combined defect indices are developed. We recommend using the Pavement Condition Index (PCI) developed in the 1970s by the United States Army Corps of Engineers (USACE) to quantify the condition of pavements on roads, streets, airports and parking areas (SHAHIN, 1994).

A Pavement Condition Index (PCI) is a numerical index that ranges from 0 (for pavements in very poor condition) to 100 (for pavements in very good condition). It can be assessed subjectively (by panels of assessors) or calculated from detailed information from a defect survey in the field. It represents the sum of the combined effects of the severity and extent of each defect that occurs on the section being analysed, and can be represented by the following equation:

$$ICP = 100 - \left[\left(\sum_i E_{i,A} \times F_{i,A}\right) + \left(\sum_i E_{i,M} \times F_{i,M}\right) + \left(\sum_i E_{i,B} \times F_{i,B}\right)\right]$$

where:

A Ei,A = extent of Defect i, with High Severity;

A Fí,A = Weighting Factor for Defect i, with High Severity.

Similarly:

A i,M = Defect i, Average Severity;

A i,B = Defect i, Low Severity.

Generally, not all defects occur simultaneously, nor do all severities for a given defect occur in a given segment in a given assessment. Based on the relative importance of each defect, Weighting Factors are defined for the high, medium and low severity levels.

The Weighting Factors differentiate the influence of the various types of defect on the loss of pavement serviceability, as some defects have a greater impact than others. They should, however, be adjusted for the various severity levels and for the operational and environmental conditions of the site being analysed.

Alternatively, the PCI can be calculated on the basis of deductible points, which gradually decrease the maximum score of 100 and which are established for each defect according to what the assessor considers to jeopardise the condition of the pavement based on the extent and severity of the defect.

The maximum deduction for each defect is set according to the relative importance of the defect.

The values of the combined indices can give an indication of which maintenance and rehabilitation strategy should be adopted. Pavement classification can use the following parameters:

- Very bad (0 to 30)
- Bad (30 to 50)
- Regular (50 to 70)
- Good (70 to 90)
- Very good (90 to 100)

Decision-making can therefore use the corresponding parameters:

- Reconstruction (0 to 30)
- Resurfacing in some cases in different ways (30 to 80)
- Maintenance appropriate to each case (80 to 100)

3.6 Need for Network Level Pavement Management Assessments

The visual situation of the pavement and the evolution of its longitudinal irregularity must be monitored throughout its useful life. This is necessary so that comprehensive decisions can be made in the long, medium and short term.

Pavement management at network level works with summarised information related to the entire road network, used for administrative decision-making. Some of the features of this system are: identifying projects that are candidates for interventions;

- Prioritise projects according to their performance, traffic, cost to users, etc;
- generate budget requirements in the short, medium and long term;
- analysing intervention strategies in view of the current condition and forecasting the future condition.

The actions to be considered are: corrective maintenance, preventive maintenance, deferred action, reinforcement, reconstruction.

Analysing the various ways of classifying and evaluating pavements, it is clear that the French technical bodies' approach to the problem has an advantage over the North American or Brazilian approaches.

In France, evaluations are carried out that make it possible to determine the problem in a broader way, defining whether the pavement being analysed has structural problems or not, which allows decisions to be made more quickly and easily. These decisions are also based on measurements of pavement deflections or irregularities. In the work by Prestes (2001), the experience of visually assessing the situation of a pavement section is described, concluding that the French method was more agile.

For project-level considerations, other information is taken into account, such as the importance of the road, financial availability and special cases.

3.7 The Need for Project-Level Pavement Management Evaluations

When thinking about 'pavement management at project level', some considerations are necessary. It is essential that accurate measurements are taken when assessing pavement conditions, so a standardised set of procedures is a must.

Also essential are: constant supervision at the time of construction, proper estimation of the pavement's useful life and knowledge of the various road rehabilitation options. This is the only way to make the right decisions for each stretch of road and design the best solutions .

Analysing the various ways of classifying and evaluating pavements, we come to the conclusion that the North American technical bodies' approach to the problem has an advantage over the French approach. The approach taken by Brazilian technical bodies is similar to that of the US, but it does present some inconsistencies and unnecessary detail.

In the US, the definitions of each defect are precise, the way to quantify them is clear and, furthermore, assessments of their extent and severity can be made in various ways. This allows for more precise measurements.

In Brazil, there are subjective, objective and combined assessment methods, which are described in the standard in a simple and clear manner. However, there are some inconsistencies and unnecessary details within the 'classification' of defects, which were mentioned in chapter 2 of this work. It should be noted that some defects are described more clearly and objectively by the DNIT (National Department of Transport Infrastructure).

According to Femandes Jr. et al. (2003), the various reviews of the work carried out within the SHRP make it a very reliable assessment method. The work of Prestes (2001) also concluded that although it takes longer, the method used by the SHRP is more reliable, obtaining more detailed and accurate results.

3.8 Treatment in Urban Areas

In this case, there are different considerations, as the occupation of these areas has different characteristics to rural roads, which are normally dealt with in pavement recovery and/or rehabilitation programmes.

In the urban area, light vehicles predominate on most roads, and speeds are lower than in rural areas. There is widespread interference from the occupations that line the roads and, in some cases, from underground constructions such as water mains and sewage outfalls ( ). This reality often makes it impossible to implement what has been designed in its entirety (NAKAHARA, 2005).

At this point it is worth pointing out that some defects that occur on motorways do not have a high incidence in urban areas, such as edge cracks or unevenness between the carriageway and the carriageway, because the lateral restraints (curbs and gutters) prevent these defects from occurring (DANIELESKI, 2004). Another important observation is with regard to patching, which may normally be necessary to repair some damage to the pavement. However, there is a high frequency of patching due to works unrelated to the use of the carriageway, such as underground works (DANIELESKI, 2004).

These factors generate doubts and cause a great deal of variation in the type of decisions to be made when defining the procedures to be carried out when rehabilitating pavements. As a result, time can be lost and the pavement management carried out on site can be affected.

For an agile and simple assessment of the problems faced in urban areas, the French method can be used efficiently. This method initially assesses whether the defect has a structural origin or not, and if there is a problem in the pavement structure, an estimate is made of the type of work to be carried out and an IS value is assigned. According to this value, which includes scores ranging from 1 to 7, it is possible to decide whether there will be a need for maintenance or reinforcement. This initial assessment can be very useful in the urban case, which has a wide variety of occurrences. The US method, on the other hand, requires a higher cost with the use of specific labour, and such detailed work is not always necessary. A more 'macro' or 'network' approach may be more appropriate for these areas.

3.9 Airport treatment

For this type of analysis, safety is much more important than comfort, since any irregularity or loss of grip could lead to major accidents, probably with serious and/or fatal casualties.
Some defects that occur at airports do not arise routinely on motorways, such as: vegetation growth in cracks or fissures, even if they are tiny, and the spillage of oil or other products. These occurrences can be much more damaging and constant than on rural roads (MAGGIE COVALT, 2006).

Surface deformation, corrugation, wear, cracks, especially fatigue cracks, and pumping, especially at the edges of the lanes, are defects that occur more frequently and require special attention (MAGGIE COVALT, 2006). To this end, it is essential to carefully limit the loads that the pavement can withstand and to strictly control drainage, as well as complying with the standards rigorously when the pavement is being built.

Evaluations need to be routine. Detailed inspections are recommended annually and general inspections monthly (TEXAS DOT, 2000). A visual survey of defects should be carried out. Pavement-tyre friction and road irregularity should also be assessed.

According to the research carried out for this work, airport pavement assessments need to be thorough and precise, using objective methods, and investments must be made before any deterioration of the asphalt concrete.

The conclusion is that any decision must pass through the 'network level' quickly and reach the 'project level', this level being detailed by the DNIT standards refined by the SHRP standards.

3.10 Conclusions on Pavement Defect Assessments

A Pavement Management system will work efficiently the more efficient the information it is subsidised with is. Therefore, combining subjective and objective evaluations can be very useful. Photos and film footage of the stretches studied can also be used. In some cases, the survey is carried out on foot (FERNANDES JR. et al., 2003), as in the development of the IGG calculation. It is important to record the extent and severity of each defect found.

According to the bibliographies analysed, the French method is agile because it takes less time to apply, and it is interesting to consider it in cases where there is great variation in pavement use and safety is not so compromised by lower traffic speeds (as is the case in urban areas) or when a macro view is required (management at network level).

The method used in the United States is more advisable when analysing pavements where safety is extremely important (such as airport pavements) or when a detailed view of a specific road section is required (project-level management).

Considering the case of Brazilian standards, according to Femandes Jr. et al. (2003), the various revisions of the work developed within the SHRP make it a more reliable assessment method than, for example, the manuals published in Portuguese (ARB, 1978 and DOMINGUES, 1993). It is worth mentioning that Domingues (1993) was one of the works that subsidised the DNIT 005/2003-TER Standard, so it is assumed that the reliability of the Brazilian standards falls short of the US standards.

Based on the assessment of the pavement's condition, the intervention to be carried out will be defined. Therefore, this data must be collected in such a way as to provide a clear assessment of the causes of the defect and the extent or scope of the defect. It is also necessary to know the existing constraints in order to adopt the appropriate solution. These constraints can be: financial, difficult traffic control, useful life of the project, geometric elements of the road, materials and equipment, quality and efficiency of the labour force, etc.

Chapter 4 presents a case study that correlates an objective assessment of a stretch of road carried out by the DER/SP/Brazil with objective and subjective assessments from the point of view of the three countries analysed. Chapter 5 proposes some adjustments to the Brazilian standards in order to make them clearer, more objective and easier to apply.

CHAPTER 4

CASE STUDY

4.1 Initial considerations

The stretch of road used for the Case Study is located on a segment of SP 253, Rodovia Deputado Cunha Bueno, between kilometre 142.000 and kilometre 173.280, i.e. between highways SP 330 and SP 255, belonging to the Ribeirão Preto regional office of DER/SP. The carriageway is 7.00 metres wide and the shoulders are paved. It has an undulating layout, characterised by horizontal curves of medium radius, interconnected by tangents. The vertical route has ramps with varying gradients. In general, the region has stable geological conditions.

The road was in poor condition, very worn, with permanent cracks and deformations, potholes of various sizes and patches along the entire stretch, both on the carriageway and on the paved shoulders, requiring rehabilitation work. The stretch for the Case Study was set between kilometre 170.300 and kilometre 173.300. This segment was considered sufficient for comparison, given that the road was in a fairly homogeneous state throughout its length.

The road condition survey was carried out by DER/SP in 2008 and the final document was issued in 2009. The stretch chosen for the case study was chosen at the end of 2010 and a full assessment of the site was carried out at the beginning of 2012, as a complement and basis for the final conclusions of the dissertation. It is worth adding that there were no major changes to the pavement conditions during this period, as the improvement works and resurfacing services for the carriageway and shoulders began in 2010, which slowed down the deterioration of the road from then on.

At the time of the survey, the pavement structure was identified as follows:

J Subgrade composed of laterite soil;

J Base made partly of gravelly soil and partly of rolled pebbles, with an average thickness of 14 cm;

J CBUQ (hot-machined bituminous concrete) road surface, between 4 and 8 cm thick.

The data collected to determine traffic at the site gave an N value (number of repetitions of the standard axle) of 4.27E+06 for 2008 and 5.04E+07 (projected) for 2018. To determine the number of vehicles, the following values were found: 1648 for passenger vehicles, 184 for public transport vehicles and 2169 for freight vehicles, totalling a VDMA (annual average daily volume) of 4002.

Below are several records of existing defects, taken from the DER/SP archive. In Figures 35 and 36, you can see an overview of the Case Study stretch, from both the right and left sides. Figures 37 and 38 show details of cracks, potholes and patches.

Figura 35 - General view of the Case Study stretch - right-hand side

Source: DER/SP archive

Figura 36 - General view of the Case Study stretch - left side

Source: DER/SP archive

Figura 37 - Large-scale patch not completely solving the problem

Source: DER/SP archive

Figura 38 - Detail of fatigue cracks with the opening of some potholes

Source: DER/SP archive

4.2 Evaluations Carried Out by DER/SP Using DNER Standards

Both the Functional Assessment and the Structural Survey were carried out by the DER/SP in order to select the maintenance and rehabilitation activities to be carried out on the road segment analysed.

4.2.1 - Functional assessments using DNER-PRO 008/94

The following types of defects were considered for assessment purposes (DNIT 005/2003):

- Type 1 - Class 1 cracks (FC-1) - FI (cracks), TTC (short transverse cracks), TTL (long transverse cracks), TLC (short longitudinal cracks), TLL (long longitudinal cracks) and TRR (shrinkage or coating cracks);
- Type 2 - Class 2 cracks (FC-2) - J (alligator cracks) and TB (edge cracks);
- Type 3 - Class 3 cracks (FC-3) - JE (interconnected cracks with alligator-type erosion) and

TBE (edge cracks with erosion);

A Type 4 - Sinking (ALP - local plastic sinking and ATP - plastic sinking in the wheel track);

A Type 5 - Ripples and Pans (O and P);

A Type 6 - Exudation (EX);

A Type 7 - Wear (D);

A Type 8 - Patches (R).

The Functional Assessment or Objective Assessment of the Surface of Flexible Pavements was carried out on the two carriageways, right and left, of the stretch under study. For this assessment, the DNER-PRO-008/94 Standard was used, which was the basis for the revised DNIT 006/2003-PRO Standard, which is part of the bibliography for this work. The truss used to measure the deflections in the wheel tracks is currently standardised by DNIT 007/2003-PRO.

The assessment carried out by the DER consisted of observing the existing defects in test stations six metres long and three and a half metres wide (the local carriageway) and determining the deflections, in millimetres, in the outer and inner wheel tracks. Note that DNIT 006/2003-PRO determines 3.00 metres before and 3.00 metres after each station.

For the stretch under study, the spacing between the evaluation surfaces (6.00 m long) was 40 metres, alternating in relation to the side of the roadway, i.e. every 80 metres on each side, right and left. The deflections in the wheel tracks were measured in millimetres at all test stations, both on the inner and outer tracks, using the metal measuring truss. It should be noted that DNIT 006/2003-PRO indicates spacing of 20 metres, alternating in relation to the side of the roadway, i.e. every 40 metres on each side. Therefore, the assessment carried out covered half as many stations as it would have done if it had strictly complied with the DNIT standard. This situation was considered sufficient due to the fact that the pavement is homogeneous across its carriageways.

The report drawn up by the DER/SP presented the results of the surveys carried out in the field, as well as the calculation of the **GGI** (Global Severity Index) for each homogeneous segment. The segment selected for the Case Study, i.e. between kilometre 170.300 and kilometre 173.300, had the following GGI values, all of which were considered to be in 'RUIM' condition:

J 170.3 to 171.3 - left side - IGG = 117.71 / 129.88 (this section covers two segments, which is why it has two IGG values)

J 172.3 to 173.3 - right side - IGG = 130.03

It should be noted that determining the GGI involves several approximations, which makes the value obtained inaccurate. Therefore, a comment that can be made in this work is that it is not necessary for GGI values to be presented to two decimal places.

4.2.2 - Structural survey using the Benkelman Beam

The Structural Survey, or Structural Assessment, was carried out using the Benkelman Beam, a device designed to measure deflections in pavements. The DNER-PRO-11/79 and DNER-PRO-269/94 standards were used to calculate the deflections and assign grades to the section analysed. The spacing between the evaluation surfaces was also 40 metres, alternating in relation to the side of the roadway, i.e. every 80 metres on each side, right and left.

The report prepared by the DER/SP presented the results of the surveys regarding the structural capacity of the motorway on this stretch. The segment selected for the Case Study showed the following results:

V 170.3 to 171.3 - left side - was classified as having 'poor or bad structural quality', with reinforcement or reconstruction of the pavement recommended;

V \T2.(3 to 173.3 - right-hand side - was classified as having 'regular structural quality' and pavement reinforcement was recommended.

4.3 Evaluations carried out in the case study

The purpose of these evaluations was to compare the results obtained by the DER/SP, when analysing the Case Study road, with the results obtained using: DNIT standards (DNIT 008/2003 and DNIT 009/2003), the Pavement Condition Index concept, and French standards.

4.3.1 - Continuous Visual Survey (Standard DNIT 008/2003-PRO)

For this assessment, the chosen stretches were measured and analysed in order to obtain the IGGE,

ICPF and IES. In the case of determining the IGGE, the DNIT 008/2003-PRO standard only takes into account the following defects: Fatigue cracks, permanent deformation, potholes and patches.

On the first segment, the left side (LE), the following values were found: 30% of the area with fatigue cracks, 25% of the area with permanent deformations, 15 (fifteen) patches, no holes. On the second segment, the right side (LD), the following values were found: 40% of the area with fatigue cracks, 25% of the area with permanent deformations, 5 (five) holes, 15 (fifteen) patches.

Codes are assigned to each frequency: A - high, M - medium, B - low. Factors are also assigned to each of these frequencies: Fpr - relative to pans and patches, Foap - relative to deformations, Ft - relative to cracks. And each defect is assigned a weight: Ppr - relative to pans and patches, Poap - relative to deformations, Pt - relative to cracks. These values are all defined in standardised tables. The tables used were taken from DNIT Standard 008/2003-PRO and can be found in Annex 3, page 109. Using these tables, the frequency (found in the field) of cracks, deformations, potholes and patches is correlated to these codes, factors and weights, as shown in Table 4.

Table 4 - Values of the elements collected in the field to calculate the IGGE, correlated to the codes, factors and weights

Evaluated segments of the stretch under study	170,300-171,300 (LE)	172,300-173,300 (LD)
Frequency of cracks	30% - code M	40% - code M
Deformation Frequency	25% - code M	25% - code M
Frequency of Pots and Patches	15* - code A	20** - code A
Ft factor	30	40
Foap factor	25	25
Factor Fpr	15	20
Pt	0,45	0,45
Poap	0,70	0,70
Ppr	1,00	1,00

* The number 15 was found by adding up the number of potholes and patches (0+15).

** The number 20 was found by adding up the number of holes and patches (5+15)

Applying the formula for calculating the IGGE:

IGGE = (Pt x Ft) + (Poap x Foap) + (Ppr x Fpr)

The following values are obtained:

1) Section 170.3 to 172.3 (left-hand side):

IGGE = (0.45 x 30) + (0.70 x 25) + (1.00 x 15) = 46

2) Section 172.3 to 173.3 (right-hand side):

IGGE = (0.45 x 40) + (0.70 x 25) + (1.00 x 20) = 57.5

To determine the ICPF, Standard DNIT-008/2003-PRG takes into account all the defects listed in Standard DNIT-005/2003-TER. The ICPF value for the Case Study can only be estimated in view of the fact that the assessor was a single person, whereas the standard suggests a minimum of two assessors. The score given, taking into account the severity and extent of the existing defects and considering the specific table for determining the ICPF through Standard DNIT-008/2003-PRG, was 2 (two).

Therefore, according to Table 3, these IGGE and ICPF values correspond to an IEF value of 5 (five), i.e. a RUIM concept. This finding is the same as that obtained by the DER in its functional assessment.

For the most accurate assessment, strict compliance with all the guidelines of DNIT 008/2003-PRO is necessary. This includes filling in all the annexes and taking careful measurements of the extent of the degradation at the site being analysed.

One positive aspect is that the LVC is not a strictly objective assessment, but combines a little of the Objective Assessment (IGGE values) with the Subjective Assessment (ICPF values). This can be very useful, as it takes into account the item 'safety' combined with the item 'comfort'.

On the other hand, a negative point is that potholes and patches are only assessed in terms of their quantity, and neither their extent nor their severity is taken into account.

4.3.2 - Evaluation using the Pavement Condition Index (PCI) concept

The purpose of this index is to quantify the condition of the pavement using an equation that utilises the length, severity level and weighting factor of the defects. A PCI value equal to 100 is attributed to a pavement in perfect condition. The equation is as follows (simplification of the equation on page 74); where Dij and fij are the extent and weighting factor of defect **i** with severity level **j:**

$$ICP = 100 - \sum_{ij} \sum_{ij} Dij \times f\,ij$$

The spreadsheet for completing this assessment must show the location, in detail, and the condition of the pavement. Deductible point values are assigned to be reduced from the maximum score of 100 for a pavement with no deterioration.

This type of concept allows for greater freedom of assessment, as the weighting values can be applied according to the situation being worked with, i.e. according to the type of road, traffic, pavement and defects that have appeared at the site being analysed. All defects can be considered with the weighting factors indicated in the bibliography cited (FERNANDES et al., 2003). All defects with equal weighting factors can be considered. Values can be assigned only to defects that occur on the stretch under consideration, with points deducted from the maximum value of 100, taking into account the greater or lesser importance and severity of each of these defects. In this case study, the third option was chosen because it is easier and more compatible to apply.

For the Fatigue Cracking and Permanent Deformation defects, the value of 13 deductible points was chosen because they are found on a large part of the pavement, but have moderate severity. A value of 1 was chosen for the Lane-Side Unevenness defect, as it was present in a small part of the chosen stretch and was of low severity. A value of 13 deductible points was chosen for the pothole and patch defects, as they are found on a large part of the road and are of high severity.

Although the pavement was aged, it was not possible to detect wear on the sections studied due to the presence of many cracks and patches along the entire length of the road.

Figure 39, taken from the DER/SP archive, illustrates a location on the stretch under construction, at the time of the traffic interruption operation in one direction, where it is possible to see a very high number of heavy vehicles, a situation that may have led to the acceleration of fatigue cracks and permanent deformation.

For the stretches studied, the ICP determination is shown in Tables 5 and 6.

Figure 39 - High occurrence of heavy vehicles on the road used in the Case Study

Source: DER/SP archive

Table 5 - ICP calculation with deductible points assigned according to severity and length: Section I (170,300 to 171,300 - LE)

DEFECTS	INTERVAL	DEDUCTIBLE POINTS
Fatigue cracks	0-20	13
Cracks in blocks	0-5	0
Edge cracks	-	0
Longitudinal cracks	0-5	0
Reflection cracks	-	0
Transverse cracks	-	0
Patches	0-15	13
Pans	0-15	0
Permanent Deformation	0-20	13
Corrugation	-	0
Exudation	-	0
Polished Aggregates	-	0
Wear and tear	0-10	0
Gap Lane Side	0-5	1
Pumping	0-5	0

Table 6 - ICP calculation with deductible points assigned according to severity and length: Section II (172,300 to 173,300 - LD)

DEFECTS	INTERVAL	DEDUCTIBLE POINTS
Fatigue cracks	0-20	13
Cracks in blocks	0-5	0
Edge cracks	-	0
Longitudinal cracks	0-5	0
Reflection cracks	-	0

Transverse cracks	-	0
Patches	0-15	13
Pans	0-15	13
Permanent Deformation	0-20	13
Corrugation	-	0
Exudation	-	0
Polished Aggregates	-	0
Wear and tear	0-10	0
Gap Lane Side	0-5	1
Pumping	0-5	0

ICP (table 5) = 100 - (13+13+13+1) = 60, for the left side;

ICP (table 6) = 100 - (13+13+13+13+1) = 47, for the right-hand side.

A value of 60 was found for Section I. A value of 47 was found for Section II. For Section II, a value of 47 was found, both requiring resurfacing, in accordance with the limits cited (FERNANDES et al., 2003):

J 0 to 30 = reconstruction;

J 30 to 80 = resurfacing;

J over 80 = appropriate maintenance for each case.

Or considered regular (left side) and bad (right side), according to the limits adopted by Lopes et al. (2008):

J 0 to 30 = very bad;

J 30 to 50 = bad;

J 50 to 70 = regular;

J 70 to 90 = good;

90 to 100 = very good.

This type of assessment has the advantage of being very agile and objective, when compared, for example, with LVC, and can help with management at network level.

4.3.3 - Subjective Assessment (Standard DNIT-009/2003-PRO)

The subjective assessment applied to the Case Study aims to correlate this analysis with the carried out in items 4.3.1 (LVC) and 4.3.2 (ICP), and also with the values found by the DER when it carried out the objective assessment of the stretch. This assessment is similar to the subjective assessment used in the USA. For this Subjective Assessment, the DNIT 009/2003 - PRO standard is used.

are grouped into just three: good, average and bad.

4.4 Comparisons between the results obtained

The work carried out by the DER/SP showed very similar results to the work carried out in the Case Study. Both the survey of defects through the Objective Assessment, based on the calculation of the IGG, and the survey of the pavement's structural state, using the Benkelman beam, indicated a segment of poor quality, in need of recovery and reinforcement work.

The surface on the right had a higher GGI value. On the other hand, the structure on the left showed greater deterioration. This work required a wide-ranging survey, taking into account all the defects in the standards, and required a series of calculations to complete.

The procedure used in section 4.3.1 took into account structural defects, according to the French standard, to determine the IGGE, i.e. fatigue cracks, permanent deformations and patches. Potholes are also included in the calculation. Therefore, the preparation of Table 4 and the IGGE calculations took less time. It should be noted that the determination of the ICPF, which evaluates all the defects indicated in Standard DNIT-005/2003-TER, could have been more precise with a greater number of evaluators. The final score (IEF) led to the same conclusion as the DER/SP survey, i.e. the need for rehabilitation and reinforcement, given the poor quality of the pavement. The Objective Assessment using the ICP (item 4.3.2) manages to cover the defects in a broad way and, at the same time, requires less and more agile work. The conclusion was the same: resurfacing is needed. The Subjective Assessment of item 4.3.3 led to the same result, emphasising that the assessment tends to show better results (lower standard deviation) the greater the number of assessors, without forgetting, however, that the greater number of assessors means higher costs.

The way of assessing using the methods used in France, item 4.3.4, which combines the agility of items 4.3.1 and 4.3.2 (LVC and ICP), was not jeopardised by the participation of a single assessor. This visual survey considers numerical values and is not based on a subjective assessment by the assessor(s) of the rolling quality, although it still depends on the 'common sense' of the 'observer'. This method can be a valuable tool for a preliminary assessment as it is very agile and easy to apply. Next, Chapter 5 offers some suggestions for improving the way Pavement Management is handled in

Brazil.

CHAPTER 5

CONCLUSIONS

There are some incoherencies and inconsistencies in Brazilian standards regarding the definition of some defects, which can lead to doubts or difficulties at the time of assessment (quantifying the extent and determining the level of severity). Some aspects could be improved by incorporating and adapting aspects of North American and French standards, with a view to defining defects clearly and simply and, consequently, obtaining more efficient Brazilian standards.

Despite its great importance, cracking is not as detailed in Brazilian standards as it is in US and French standards. French standards, for example, distinguish between fatigue cracks and other cracks. Fatigue cracks are considered structural defects, while the others are considered defects of other than structural origin (block, edge, transverse or longitudinal cracks). Longitudinal cracks, when they appear in the wheel track, may have a structural origin. It should also be pointed out that the subdivision of fatigue cracks (ageing and reflection cracks) presented in the DNIT Manual is confusing and changes concepts that are already recognised as appropriate.

Grouping Corrugation and Slippage defects together, as the foreign standards studied do, could make the Brazilian standards clearer and simpler, making it easier to identify the type and level of severity of the defect and, consequently, the appropriate selection of the maintenance and rehabilitation activity to be carried out.

For the French standard, potholes can have a variety of origins and must be corrected immediately, while a patch is classified as a structural defect. The SHRP programme manual also distinguishes between patches and potholes, so much so that the test sections were monitored continuously, with patches and potholes recorded even when the potholes were corrected immediately.

The Brazilian standards for calculating the IGGE (Standard DNIT-008/2003-PRO) group together the defects Potholes and Patches, as presented in the Case Study, which can jeopardise the assessment, as a patch is not always made to cover a pothole. It is important to emphasise that the Continuous Visual Survey, which does not assess the extent of patches and potholes, but only their quantity, can lead to distortions as it does not assess severity either.

The definitions of the three standards complement each other, so the DNIT Manual would present clearer and more complete definitions if it combined the definitions of the three countries, with the appropriate adaptations suggested in this work, to eliminate the inconsistencies and complexities that currently exist in the national standards.

By analysing how pavement defects arise and evolve, it is of great importance for pavement management in Brazil to classify and evaluate pavements by differentiating between structural and non-structural defects. Differentiating treatment, as the French standards do, makes it easier to survey defects and aids decision-making. Knowing whether or not a defect is caused by structural problems can prevent expensive and unnecessary structural assessments and even help prevent its occurrence in the future, by helping to predict where it is most likely to occur.

The initial survey, at network ('macro') level, is capable of defining whether a given road section has structural defects or not. It is the starting point and the main form of assessment for urban pavement management.

With regard to pavement management at project level, the North American standards are objective and precise, which is also the case when it comes to airport pavements. They should therefore be a reference for the DNIT Manual when it comes to determining the severity and extent of defects.

On the other hand, the French standards have the inconvenience of using some terms, such as 'clearly open cracks', which leave room for different interpretations by different assessors regarding the severity of the defects. It is worth pointing out here that the levels of severity of the defect Gap between carriageway and shoulder are only mentioned in the French standards, and these levels should be taken into account for a future general classification in Brazil.

The determination of a Pavement Condition Index (PCI) makes it possible to define the most appropriate intervention strategy with considerable objectivity, using a decision tree. When determining the deductible points used to calculate the PCI, higher value ranges are defined for structural defects, i.e. fatigue cracks and permanent deformation of the wheel tracks. A high value is also assigned to patches (also considered a structural defect) and potholes. Although wear was not identified in the Case Study, despite having preceded the occurrence of many of the defects found, its value range should also be high.

After the initial assessment of the entire network, the frequency of new assessments should be greater for pavements subjected to heavier loads and more intense flow, suggesting annual defect surveys. On the other hand, roads with light traffic and no evidence of structural problems can be assessed every three years.

Therefore, the revision and reorganisation of Brazilian standards could take into account the following points in particular:

I will go into more detail about cracking, with emphasis on fatigue cracking, as a defect in the pavement structure (longitudinal cracking can also be classified as a structural defect when it appears in the wheel track);

J to exclude the subdivision of fatigue cracks, ageing cracks and reflection cracks from the DNIT Manual, as it mistakenly modifies concepts that are already recognised;

J groups the corrugation and slip defects into a single defect;

J deals in particular with the Patch defect, which is also a surface conservation defect, but which is nonetheless an indication of a location that will always deserve attention when evaluations are carried out;

J used the Panela (Hole) defect in the evaluations because it still occurs so frequently on Brazilian pavements;

J consider the severity levels and extent (area or length) in all cases when assessing and quantifying;

J complements the definitions of defects in Brazilian standards with those used in other countries;

J classify defects by separating defects of structural origin from defects of various origins ; consider different assessments for different conditions and different locations, with the US standards being more appropriate for airports and for project-level assessments, and the French standards for urban pavements and for network-level assessments;

Use the ICP concept.

These considerations could result in a more agile, simple and efficient pavement condition assessment method, ultimately providing safer, more comfortable and more economical motorway, urban and airport pavements, which is what motivated this research project.

CHAPTER 6

BIBLIOGRAPHICAL REFERENCES

ARB (1978). **Catalogue of Pavement Surfacing Defects.** Brazilian Road Association. Translated by Hugo Alves Pequeno. São Paulo

AUTRET, P.; BROUSSE, J.L. (2001). **Vizir, A computer-supported method for estimating the maintenance needs of a road network.** Laboratoires Central des Ponts et Chaussées, LCPC.

BERNUCCI, L.B.; MOTTA, L.M.G.; CERRATI, J.A.P.; SOARES, J.B. (2007). **Asphalt Paving, Basic Training for Engineers.** Brazilian Association of Asphalt Distributors, ABEDA.

CAREY, W.N.; IRICK, P.E. (1960). **The Pavement Serviceability-Performance Concept.** Highway Research Board Bulletin 250, pp. 40-58.

DANIELESKI, M.L. (2004). **Proposal for a Methodology for the Surface Evaluation of Urban Pavements.** Conclusion Paper for the Professionalising Master's Degree in Engineering as a partial requirement for obtaining the Master's Degree in Engineering.

DER (2008). **Technical Document, Pavement Structural Assessment Report, km 142.000 to km 173.820 of SP 253.**

DER (2008). **Technical Document, Pavement Functional Assessment Report, km 142.000 to km 173.820 of SP 253.**

DNER (1994). **Objective Surface Evaluation of Flexible and Semi-rigid Pavements - Procedure Standard DNER-PRO 08 / 1994.** National Department of Transport Infrastructure. Rio de Janeiro, RJ.

DNER-M3 024/94 (1994). **Pavement - determination of deflections by Benkelman beam -** DNER Test Method.

DNIT (2003). **Defects in flexible and semi-rigid pavements - Terminology NORMA DNIT 005 / 2003 - TER.** National Department of Transport Infrastructure. Rio de Janeiro, RJ.

DNIT (2003) **Objective surface evaluation of flexible and semi-rigid pavements - Procedure NORMA DNIT 006 / 2003 - PRO.** National Department of Transport Infrastructure. Rio de Janeiro, RJ.

DNIT (2003) **Survey to assess the surface condition of homogeneous road sections with flexible and semi-rigid pavements for pavement management and studies and projects - Procedure NORMA DNIT 007 / 2003 - PRO.** National Department of Transport Infrastructure. Rio de Janeiro, RJ.

DNIT (2003) **Continuous visual survey to evaluate the surface of flexible and semi-rigid pavements - Procedure NORMA DNIT 008 / 2003 - PRO.** National Department of Transport Infrastructure. Rio de Janeiro, RJ.

DNIT (2003) **Subjective surface evaluation of flexible and semi-rigid pavements - Procedure NORMA DNIT 009 / 2003 - PRO.** National Department of Transport Infrastructure. Rio de Janeiro, RJ.

DNIT (2006). **Road Pavement Restoration Manual,** Publication IPR-720, Ministry of Transport, DNIT.

DOMINGUES, F.A.A. (1993). **MID - Manual for Identifying Defects in Asphalt Pavement Surfacings.** São Paulo, S.P.

FERNANDES JÚNIOR, J.L.; ODA, S.; ZERBINI, L.F. (2003). **Defects and Maintenance and Rehabilitation Activities in Asphalt Pavements**, University of São Paulo, São Carlos School of Engineering, Department of Transport.

FHWA (1993) - U.S. Department of Transportation. **Distress Identification Manual for the Long-Term Pavement Performance Programme.** The Strategic Highway Research Programme. National Academy of Science. Washington, D.C.

HAAS, R.; HUDSON, W.R.; ZANIEWSKY, J. (1994). **Modern Pavement Management.** Krieger Publishing Company, Malabar, Florida.

LABORATOIRE CENTRAL DES PONTS EL CHAUSSÉES, MÉTHODE D'ESSAI N°52, COMPLÉMENTE À LA MÉTHODE D'ESSAI N°38-2 (1998). **Catalogue of Surface Degradations of Chaussées.** Techniques and methods of bridge and tramway laboratories.

LOIOLA, P.R.R. (2009). **Study of alternative aggregates and binders for use in road surface treatments.** Master's dissertation. University of Ceará, Master's Programme in Transport Engineering, Fortaleza.

LOPES, Simone Becker; PFAFFENBICHLER, Paul; EMBERGER, Gunter; RIEDL, Leopold (2008). **Urban Pavement Management Using Dynamic System Modelling Directly Connected to a GIS.** Panorama Nacional da Pesquisa em Transportes 2008 / XXII Congresso Nacional de Pesquisa e Ensino em Transportes. Fortaleza-CE, 03 to 07 November 2008.

MAGGIE COVALT, P.E. (2006). **Washington State Airport Pavement Management System.** Applied Pavement Technology.

NAKAHARA, S.M. (2005). **Study of the Performance of Asphalt Pavement Reinforcements on an Urban Road Subject to Heavy Commercial Traffic**, Civil Engineer, PUC / RS, Porto Alegre, Master's Degree in Civil Engineering, UFRGS, Porto Alegre. Thesis presented to the Polytechnic School of USP for the degree of Doctor of Engineering, Area of Concentration Transport Engineering.

PRESTES, M.P. (2001). **Visual Evaluation Methods for Flexible Pavements, A Comparative Study,** Federal University of Rio Grande do Sul, School of Engineering, Professional Master's Degree in Engineering.

ROBERTS, F.L.; KANDHAL, P.S.; BROWN, E.R.; LEE, D.; KENNEDY, T.W. (1991). **Hot Mix Asphalt Materials, Mixtures Design and Construction, First Edition,** National Center for Asphalt Technology - Aubum University - Alabama, Dah-Yinn Lee - Iowa State University - Ames - Iowa, Thomas W. Kennedy - University of Texas - Austin - Texas, NAPA Education Foundation, Lanham, Maryland.

SHAHIN, M.Y. (1994). **Pavement Management for Airports, Roads and Parking Lots.** Chapman & Hall, New York.

THE AVIATION DIVISION FROM TEXAS DEPARTMENT OF TRANSPORTATION - TEXAS DOT (2000). **Pavement Management Programme for General Aviaton Airports.** Austin, Texas.

ANNEXES

Quadro resumo dos defeitos – Codificação e Classificação

FENDAS				CODIFICAÇÃO	CLASSE DAS FENDAS		
Fissuras				FI	-	-	-
Trincas no revestimento geradas por deformação permanente excessiva e/ou decorrentes do fenômeno de fadiga	**Trincas Isoladas**	Transversais	Curtas	TTC	FC-1	FC-2	FC-3
			Longas	TTL	FC-1	FC-2	FC-3
		Longitudinais	Curtas	TLC	FC-1	FC-2	FC-3
			Longas	TLL	FC-1	FC-2	FC-3
	Trincas Interligadas	"Jacaré"	Sem erosão acentuada nas bordas das trincas	J	-	FC-2	-
			Com erosão acentuada nas bordas das trincas	JE	-	-	FC-3
Trincas no revestimento não atribuídas ao fenômeno de fadiga	**Trincas Isoladas**	Devido à retração térmica ou dissecação da base (solo-cimento) ou do revestimento		TRR	FC-1	FC-2	FC-3
	Trincas Interligadas	"Bloco"	Sem erosão acentuada nas bordas das trincas	TB	-	FC-2	-
			Com erosão acentuada nas bordas das trincas	TBE	-		FC-3

OUTROS DEFEITOS				CODIFICAÇÃO
Afundamento	**Plástico**	**Local**	Devido à fluência plástica de uma ou mais camadas do pavimento ou do subleito	ALP
		da Trilha	Devido à fluência plástica de uma ou mais camadas do pavimento ou do subleito	ATP
	De Consolidação	**Local**	Devido à consolidação diferencial ocorrente em camadas do pavimento ou do subleito	ALC
		da Trilha	Devido à consolidação diferencial ocorrente em camadas do pavimento ou do subleito	ATC
Ondulação/Corrugação - Ondulações transversais causadas por instabilidade da mistura betuminosa constituinte do revestimento ou da base				O
Escorregamento (do revestimento betuminoso)				E
Exsudação do ligante betuminoso no revestimento				EX
Desgaste acentuado na superfície do revestimento				D
"Panelas" ou buracos decorrentes da desagregação do revestimento e às vezes de camadas inferiores				P
Remendos		**Remendo Superficial**		RS
		Remendo Profundo		RP

NOTA 1: Classe das trincas isoladas

FC-1: são trincas com abertura superior à das fissuras e menores que 1,0mm.

FC-2: são trincas com abertura superior a 1,0mm e sem erosão nas bordas.

FC-3: são trincas com abertura superior a 1,0mm e com erosão nas bordas.

NOTA 2: Classe das trincas interligadas

As trincas interligadas são classificadas como FC-3 e FC-2 caso apresentem ou não erosão nas bordas.

ANNEX 1 - Classification of Defects

Standard DNIT 005/2003-TER

Valor do Fator de Ponderação

Ocorrência Tipo	Codificação de ocorrências de acordo com a Norma DNIT 005/2002-TER "Defeitos nos pavimentos flexíveis e semi-rígidos – Terminologia" (ver item 6.4 e Anexo D)	Fator de Ponderação fp
1	Fissuras e Trincas Isoladas (FI, TTC, TTL, TLC, TLL e TRR)	0,2
2	FC-2 (J e TB)	0,5
3	FC-3 (JE e TBE) NOTA:Para efeito de ponderação quando em uma mesma estação forem constatadas ocorrências tipos 1, 2 e 3, só considerar as do tipo 3 para o cálculo da freqüência relativa em percentagem (fr) e Índice de Gravidade Individual (IGI); do mesmo modo, quando forem verificadas ocorrências tipos 1 e 2 em uma mesma estação, só considerar as do tipo 2.	0,8
4	ALP, ATP e ALC, ATC	0,9
5	O, P, E	1,0
6	EX	0,5
7	D	0,3
8	R	0,6

Índice de gravidade global (IGG)

O Índice de Gravidade Global (IGG) é obtido por meio da fórmula:

$$IGG = \sum IGI$$

onde:

Σ IGI - somatório dos Índices de Gravidade Individuais, calculados de acordo com o estabelecido no item 7.3. O Índice de Gravidade Global deve ser calculado para cada trecho homogêneo (ver Anexo C).

Conceito de degradação do pavimento

Com a finalidade de conferir ao pavimento inventariado um conceito que retrate o grau de degradação atingido, é definida a correspondência apresentada na Tabela 2 a seguir:

Conceitos de degradação do pavimento em função do IGG

Conceitos	Limites
Ótimo	0 < IGG ≤ 20
Bom	20 < IGG ≤ 40
Regular	40 < IGG ≤ 80
Ruim	80 < IGG ≤ 160
Péssimo	IGG > 160

ANNEX 2 - Weighting Factors and Pavement Degradation Concepts according to IGG - Standard DNIT 006/2003-PRG

CONCEITO	DESCRIÇÃO	ICPF
Ótimo	NECESSITA APENAS DE CONSERVAÇÃO ROTINEIRA	5 - 4
Bom	APLICAÇÃO DE LAMA ASFÁLTICA - Desgaste superficial, trincas não muito severas em áreas não muito extensas	4 -3
Regular	CORREÇÃO DE PONTOS LOCALIZADOS OU RECAPEAMENTO - pavimento trincado, com "panelas" e remendos pouco freqüentes e com irregularidade longitudinal ou transversal.	3 - 2
Ruim	RECAPEAMENTO COM CORREÇÕES PRÉVIAS - defeitos generalizados com correções prévias em áreas localizadas - remendos superficiais ou profundos.	2 - 1
Péssimo	RECONSTRUÇÃO - defeitos generalizados com correções prévias em toda a extensão. Degradação do revestimento e das demais camadas - infiltração de água e descompactação da base	1 - 0

Panelas (P) e Remendos (R)		
FREQÜÊNCIA	Fator Fpr Quantidade/Km	GRAVIDADE
A - ALTA	≥ 5	3
M - MÉDIA	2 - 5	2
B - BAIXA	≤ 2	1
Demais defeitos (trincas, deformações)		
FREQÜÊNCIA	Fatores Ft e Foap (%)	GRAVIDADE
A - ALTA	≥ 50	3
M - MÉDIA	50 - 10	2
B - BAIXA	≤ 10	1

Pesos para cálculo

GRAVIDADE	Pt	Poap	Ppr
3	0,65	1,00	1,00
2	0,45	0,70	0,80
1	0,30	0,60	0,70

Freqüência de defeitos

Panelas (P) e Remendos (R)		
Código	Freqüência	Quant./km
A	Alta	≥ 5
M	Média	2 –5
B	Baixa	≤ 2
Demais defeitos		
Código	Freqüência	% por km
A	Alta	≥ 50
M	Média	50 – 10
B	Baixa	≤ 10

ANNEX 3 - Gravities and Frequencies, Calculation Weights, ICPF Concepts and Descriptions - Standard DNIT 008/2003-PRO

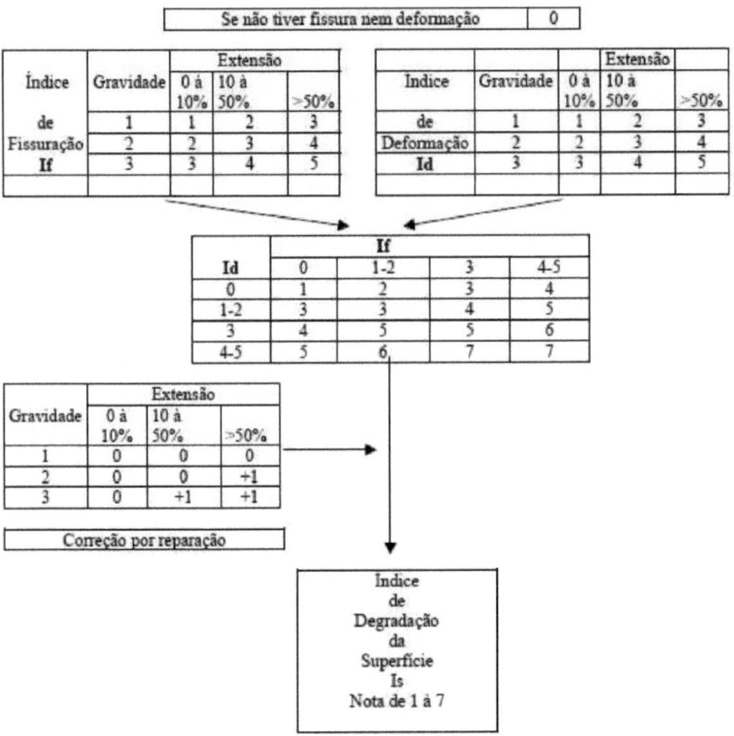

DETERMINAÇÃO DO ÍNDICE DE DEGRADAÇÃO

Se não tiver fissura nem deformação	0

Índice de Fissuração If	Gravidade	Extensão 0 à 10%	10 à 50%	>50%
	1	1	2	3
	2	2	3	4
	3	3	4	5

Indice de Deformação Id	Gravidade	Extensão 0 à 10%	10 à 50%	>50%
	1	1	2	3
	2	2	3	4
	3	3	4	5

Id \ If	0	1-2	3	4-5
0	1	2	3	4
1-2	3	3	4	5
3	4	5	5	6
4-5	5	6	7	7

Gravidade	Extensão 0 à 10%	10 à 50%	>50%
1	0	0	0
2	0	0	+1
3	0	+1	+1

Correção por reparação

Índice de Degradação da Superfície Is Nota de 1 à 7

ANNEX 4 - Determination of IF, ID, repair correction VIZIR method

MIX
Papier aus verantwortungsvollen Quellen
Paper from responsible sources
FSC® C105338

Printed by Books on Demand GmbH, Norderstedt / Germany